THE LITTLE BOOK OF SCIENTIFIC PRINCIPLES, THEORIES, & THINGS

SURENDRA VERMA

NH
NEW HOLLAND

This edition first published in 2006 by
New Holland Publishers (UK) Ltd
London • Cape Town • Sydney • Auckland

Garfield House, 86–88 Edgware Road, London W2 2EA, UK
www.newhollandpublishers.com

80 McKenzie Street Cape Town 8001 South Africa
14 Aquatic Drive Frenchs Forest NSW 2086 Australia
218 Lake Road Northcote Auckland New Zealand

First published in Australia in 2005 by
Reed New Holland
an imprint of New Holland Publishers (Australia) Pty Ltd

10 9 8 7 6 5 4 3 2 1

ISBN 1-84537-527-0

Publisher: Louise Egerton
Project Editor: Yani Silvana
Designer: saso content and design pty ltd
Production: Linda Bottari
Printed and bound in India by Replika Press Pvt. Ltd.

How To Use This Book

This book introduces more than 175 laws, principles, theories, hypotheses, rules, postulates, theorems, experiments, models, systems, paradoxes, equations, constants, and other ideas that form the foundations of science. (The meanings of the terms law, principle etc. are explained in the appendix under "The Scientific Method," on p. 202.)

The entries are arranged in chronological order. The popular name of the idea is used; in most cases it includes the name of the person who formulated the idea, for example, Archimedes' Principle.

You may read the book in order, or dip into it at random. Each entry is complete in itself, but cross-references are provided where appropriate. The key on the opposite page explains the order of information in each entry.

The comprehensive index will help you to find the ideas covered in this book. Main entries appear in bold in the index, as do some additional terms/ideas (also bolded in text).

Abbreviations used

- **c.** circa (meaning approximately)
- **b.** born
- **p.** page
- **pp.** pages

The year of discovery and the country of discovery (if this is not the country of birth of the discoverer, the country is mentioned in the entry).

The name(s) of the discoverer(s), their years of birth and death (if any) in brackets.

The popular name of the scientific idea.

1621

HOLLAND

Snell's Law

Willebrord Snell (1580–1626)

The idea in brief.

During refraction of light, the ratio of the sines of the angles of incidence (*i*) and refraction (*r*) is a constant equal to the refractive index of the medium.

Explanation of the idea or its mathematical representation, and, where relevant, the importance of the idea.

Useful and interesting information about the idea, its discoverer(s), its impact and/or applications.

In equation form, $n_1 \sin i = n_2 \sin r$, where n_1 and n_2 are the respective refractive indices of the two media. The refractive index of a substance is a measure of its ability to bend light. The higher the number the better light is refracted. For example, the refractive index of diamond, 2.42, is the highest of all gems.

Refraction is the change in direction of a ray of light when it crosses the boundary between the two media. It happens because light has different speeds in different media. A ray of light entering a medium where the speed of light is less (from air to water, for example) bends toward the perpendicular to the boundary of the two media (*see* diagram). It bends away from the perpendicular when it crosses from water to air. Refraction was known to ancient Greeks, but Snell, a Dutch mathematician, was the first to study it. Refraction is responsible for many optical illusions such as a stick in water appearing to be bent.

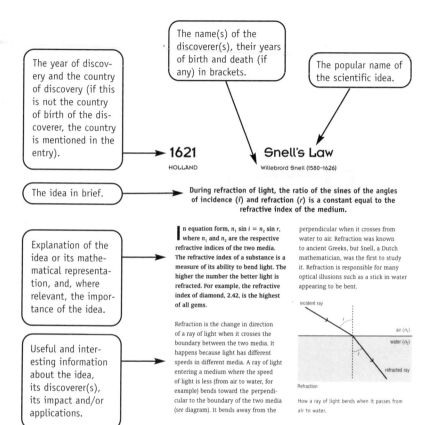

incident ray

air (n_1)

water (n_2)

refracted ray

Refraction

How a ray of light bends when it passes from air to water.

25

5

Pythagorean Theorem

Pythagoras (c. 580 – c. 500 BC)

In a right-angled triangle, the square on the hypotenuse is the sum of the squares on the other two sides.

The theorem can also be written as a general law: $a^2 + b^2 = c^2$, where c is the length of the hypotenuse of a right-angled triangle, and a and b the lengths of other two sides. Pythagoras' theorem is a starting point for trigonometry, which has many practical applications such as calculating the height of mountains and measuring distances.

Legend has it that Pythagoras was once walking on the checkered floor of a temple in Egypt. The floor had alternatively colored squares. The shadow of pillars was falling obliquely across these squares. The shadows and squares suggested different geometrical patterns. His interest in geometry led him to study these patterns from different angles and then to the discovery of the proof of the theorem.

Pythagoras was the first to prove the relationship between the sides of a right-angled triangle, but he did not discover it—it was known to Babylonians for nearly 1000 years before him. He was also the first to discover that the Earth was a sphere. This discovery was confirmed by Eratosthenes (p. 15).

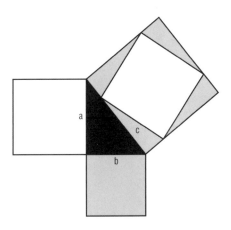

The square on the hypotenuse of the right-angled black triangle is equal to the sum of the squares on the other two sides. (The white squares are the same size; the four shaded pieces make the shaded square.)

Zeno s Paradoxes

Zeno (c. 495–c. 430 BC)

Motion is an illusion.

Zeno devised four paradoxes, all of which seemed to prove the impossibility of motion.

The most famous of Zeno's paradoxes is that of Achilles and the tortoise: the faster of two runners can never overtake the slower, if the slower is given any start at all. Suppose Achilles, a hero of the Trojan War, reputed to be the swiftest runner ever known, can run ten times as fast as a tortoise and the tortoise has a 100-yard head start. When Achilles has run 100 yards the tortoise will have crawled 10 yards, and so will be 10 yards in front, and so on. Mathematically, Achilles can only keep getting nearer and nearer to the tortoise, but he can never overtake it.

Zeno's paradoxes were based on the false assumption that space and time are infinitely divisible; that is, the sum of an infinite number of numbers is always infinite. Though they were based on fallacies, the paradoxes remained unsolved for two millennia. In the seventeenth century AD the Scottish mathematician James Gregory showed that an infinite number of numbers can add up to a finite number. Such a series of numbers is called the convergence series, which occurs when the difference between each number and the one following it becomes smaller throughout the sequence. The race between Achilles and the tortoise is not made up of a number of tiny distances but it is continuous until the end. Of course, Achilles will win the race.

Democritus' Atomic Theory

Democritus (c. 460–c. 370 BC)

Matter is made up of empty space and an infinite number of tiny invisible particles called *atomos* or atoms.

Democritus' atomic theory was probably based on previous ideas of other Greek philosophers. It was the first scientific attempt to explain the nature of matter; however, many of Democritus' assumptions have now been proved wrong.

Democritus, probably the greatest of the early Greek physical philosophers, also suggested that atoms could not be divided indefinitely into smaller parts, and that it was impossible to create new matter. He said that atoms were always in motion and as they moved about they collided with other atoms: sometimes they interlocked and held together, sometimes they rebounded from collisions. The Roman poet Lucretius (c. 99–c. 55 BC) imagined Democritus' atoms with hooks that fastened them together.

Very little is known about Democritus' life, but we know all about his atomic theory from the second-century AD Greek biographer Diogenes Laertius' book *Lives of Eminent Philosophers*. Laertius lists 73 books written by Democritus, of which only fragments have survived. Democritus is believed to have said: "I would rather discover one scientific fact than become King of Persia."

The great Greek philosopher Aristotle (384–322 BC) rejected Democritus' idea of the atom and said that the matter was completely uniform and continuous. The influence of Aristotle was extraordinary. His concept of matter was basically wrong, but it was accepted for some 20 centuries until it was replaced by Dalton's atomic theory (p. 65) in 1808.

Hippocratic Corpus

Hippocrates (c. 460–c. 377 BC)

About 60 ancient medical extant writings, known collectively as the Hippocratic corpus.

The corpus is the oldest surviving Western scientific text. It laid the foundations of the Western medical tradition. Although its remedies are now considered "imaginative," the corpus speaks the language of science; it does not speak of spells, demons, or gods.

Hippocrates (known as Buqrat in the Muslim world) is now remembered as the Father of Medicine, but very little is known about him. He was a contemporary of Socrates and lived on the island of Cos. Celsus, a medical encyclopedist of the first century AD, describes him as *primus ex omnibus memoria dignis* (the first physician worthy of being remembered).

Hippocratic medicine was based on the balance of four elements—water (cold and moist), air (moist and hot), fire (hot and dry), and earth (cold and dry)—and four humors (bodily fluids)—phlegm, blood, bile, and black bile. Sickness was the sign of imbalance. If, for example, the sickness consisted of an excess of cold, moist humors, then the task of the physician was to restore the balance. Physicians no longer practice Hippocratic medicine, but his name survives in the Hippocratic oath that medical students take on graduation day in many medical schools. Hippocrates required his pupils to take an oath to practice medicine according to certain ethical rules.

Euclid's Postulates

Euclid (lived c. 325–c. 265 BC)

(1) A straight line can be drawn between two points.
(2) A straight line can be extended indefinitely in either direction.
(3) A circle can be drawn with any given center and radius.
(4) All right angles are equal.
(5) If two lines are drawn which intersect a third in such a way that
the sum of the inner angles on one side is less than two right angles,
then the two lines will eventually meet (or, parallel lines never meet).

These five postulates form the basis of Euclidean geometry. Many mathematicians do not consider the fifth postulate (or the parallel postulate) as a true postulate, but rather as a theorem that can be derived from the first four postulates. Aspects of Euclidean geometry are still taught in schools.

Euclid's *Elements* is one of the most widely read textbooks of all time. It was the standard textbook on geometry until other kinds of geometries, such as Cartesian coordinate geometry, were discovered in the seventeenth century. Virtually nothing is known about Euclid's life. He studied in Athens and then worked in Alexandria during the reign of Ptolemy I. Two well-known anecdotes are associated with him. Ptolemy asked Euclid if there was an easier way to learn geometry than by studying all the theorems. Euclid replied: "There is no royal road to geometry." According to another anecdote, one of his students complained, as students often do, that learning geometry was pointless—it had no practical value. Euclid ordered a slave to give the student a coin so that he could make a profit in studying geometry.

Elements begins with 23 definitions (such as point, line, circle, and right angle), five postulates, and five "common notions." From these foundations Euclid proved 465 theorems. A postulate (or axiom) claims something is true or is the basis for an argument. A theorem is a proven proposition, which is a statement with logical constraints. Euclid's common notions are not about geometry; they are elegant assertions of logic:

1. Two things which are both equal to a third thing are also equal to each other.

2. If equals are added to equals, the wholes are equal.

3. If equals are subtracted from equals, the remainders are equal.

4. Things which coincide with one another are equal to one another.

5. The whole is greater than the part.

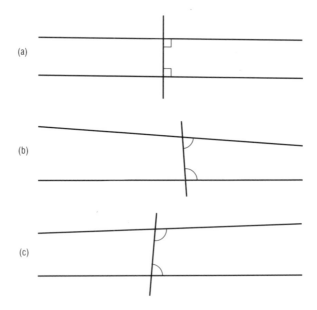

Euclid's parallel postulate: parallel lines never meet (a), but if the sum of the inner angles is less than two right angles on one side, then the lines will eventually meet (b and c).

Archimedes' Principle

Archimedes (c. 287–212 BC)

A body fully or partially immersed in a fluid is buoyed up by a force equal to the weight of the fluid displaced by the body.

The upthrust (the upward force) on a floating object, such as ship, is the same as the weight of water it displaces.

Syracuse's King Hieron, suspecting that the goldsmith had not made his crown of pure gold as instructed, asked Archimedes to find out the truth without damaging the crown. One day, while at the public baths, Archimedes noticed that the deeper he descended into the tub, the more water flowed over the edge. He suddenly realized that he could now solve the problem of the king's crown.

He was so excited by his discovery that he ran naked through the street, shouting "Eureka! Eureka!," a Greek word which means "I have found it."

Archimedes, the greatest scientist and mathematician of antiquity, immersed the crown in a container brimful of water and collected the water that overflowed. When he placed a lump of pure gold equal in weight to the crown in the water, a lesser amount of water overflowed. Thus Archimedes concluded that the goldsmith had substituted some gold with a metal of lesser density such as silver.

π

Archimedes (c. 287–212 BC)

All circles are similar and the ratio of the circumference to the diameter of a circle is always the same number, known as the constant π (pi).

Like the square and cube roots of 2, π is an irrational number: it takes a never-ending string of digits to express π as a number. It is impossible to find the exact value of π; however, the value can be calculated to any required degree of accuracy.

The earliest reference to the ratio of the circumference to the diameter of a circle is in an Egyptian papyrus written in 1650 BC, but Archimedes first calculated the most accurate value. His value lies between $3\frac{1}{7}$ and $3\frac{10}{71}$, or between 3.142 and 3.141, and is accurate to two decimal places. In the eighteenth century AD the Swiss mathematician Leonhard Euler first used the letter π, the first letter of the Greek word for perimeter, to represent this ratio.

Even for designing a satellite we need know the value of π for only a few decimal places, but some mathematicians love to calculate the value of π. In 2002 Yasumasa Kanada (b. 1948) of Tokyo University used a supercomputer for the job. It took the computer, with a memory of 1024 gigabytes, 602 hours to compute the value to 124,100,000,000 decimal places.

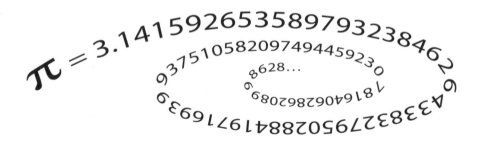

$$\pi = 3.14159265358979323846264338327950288419716939937510582097494459230781640628620899862803482534211706798214808651328230664709384460955058223172535940812848111745028410270193852110555964462294895493038196 4428810975665933446128475648233786783165271201909145648566923460348610454326648213393607260249141273724587006606315588174881520920962829254091715364367892590360011330530548820466521384146951941511609...$$

Eratosthenes' Measurement of the Earth

Eratosthenes (c. 276–194 BC)

At noon on the day of summer solstice, the sun is directly overhead in Syene (now Aswan) and there is no shadow, but at the same time in Alexandria the sun is at an angle and there is a measurable shadow.

E ratosthenes used this simple concept to calculate the circumference of the Earth.

Eratosthenes reasoned that the surface of the Earth was curved, resulting in the angle of the sun's rays being different in different locations. With the aid of his simple geometrical instruments he found that in Alexandria at noon the sun's rays were falling at an angle of 7.2 degrees, which is one-fiftieth of 360 degrees. Knowing the distance between the two places, he calculated that the circumference of the Earth was 50 times that distance. Eratosthenes' value comes to 24,451 miles, as against a true average length of 24,875 miles. An astonishing achievement!

Eratosthenes was a versatile scholar: an astronomer, mathematician, geographer, historian, literary critic, and poet. He was nicknamed "Beta" (the second letter of the Greek alphabet) because he was considered the second-best at everything.

How Eratosthenes measured the size of the Earth (diagram not drawn to scale).

Ptolemy's Earth-centered Universe

Claudius Ptolemy (c. 100–170)

The Earth is at the center of all the cosmos.

This erroneous belief dominated astronomy for 14 centuries.

"The Earth does not rotate; otherwise objects will fling off its surface like mud from a spinning wheel. It remains at center of things because this is its natural place—it has no tendency to go either one way or the other. Around it and in successively larger spheres revolve the moon, Mercury, Venus, the sun, Mars, Jupiter, and Saturn, all of them deriving their motion from the immense and outermost spheres of fixed stars," Ptolemy writes in his book, *Almagest*, in which he synthesized the work of his predecessors.

Almagest's eminence, importance, and influence can only be compared with Euclid's *Elements* (p. 11). A major part of *Almagest* (Arabic for "the greatest") deals with the mathematics of planetary motion. Ptolemy explained the wanderings of the planets by a complicated system of cycles and epicycles, which harassed astronomers for centuries. "If the Lord Almighty had consulted me before embarking upon the Creation, I should have recommended something simpler," commented Alfonso the Wise, the thirteenth-century Spanish king of Castile, and a great patron of astronomy. Ptolemy's theory was challenged by Copernicus (p. 19) and demolished by Kepler (p. 23). Ptolemy supported Eratosthenes' view that the Earth is spherical (p. 15), which encouraged Columbus in his voyages of discovery.

1202 Fibonacci Numbers

ITALY   Leonardo Pisano (aka Fibonacci) (c. 1170–c. 1250)

A series of numbers in which each successive term is the sum of the preceding two. For example: 1, 1, 2, 3, 5, 8, 13, 21, 34, 55, 89, 144 ...

The series is known as the Fibonacci sequence and the numbers themselves as the Fibonacci numbers.

The Fibonacci sequence has many other interesting mathematical properties. For example, the ratio of successive terms (larger to smaller; 1/1, 2/1, 3/2, 5/3, 8/5 ...) approaches the number 1.618. This ratio is known as the **golden ratio** and is denoted by the Greek letter phi (F). Phi was known to ancient Greeks, and Greek architects used the ratio 1: F as part of their design, the most famous of which is the Parthenon in Athens. Curiously, phi also appears in the natural world. Flowers often have a Fibonacci number of petals (look at the arrangement of florets on a cauliflower). The seeds on a sunflower are arranged in two sets of spirals. The ratio of the number of seeds in the two spirals is phi—and so is the ratio of your height to the distance from your belly button to the tip of your feet.

During his travels in North Africa, Fibonacci learned of the decimal system of numbers that had evolved in India and had been taken up by the Arabs. In his book *Liber Abaci* (1202) he introduced to Europe the Arabic numerals that we use today.

Ockham's Razor

William of Ockham (1285–1349)

Entities should not be multiplied unnecessarily.

This guiding principle in developing scientific ideas insists that you should prefer the simplest explanation to fit the facts.

William, a philosopher and theologian, came from Ockham, a village about 25 miles southwest of London. In his youth he joined the Franciscan order and studied at Oxford, where he lectured from 1317 to 1319. At Oxford, which was then a great Franciscan center of learning, William became the leader of a school of philosophy called nominalism.

He is now most remembered for his rule, known as Ockham's (also spelled Occam's) Razor, that is of vital importance in the philosophy of science. The rule has been interpreted in modern times to mean that when you have two competing theories which make exactly the same predictions, the one that is simpler is the better. In other words, the explanation requiring the fewest assumptions is most likely to be correct. Advice to computer programmers to keep their programs simple—keep it simple, stupid (KISS)—is in similar vein. But we must also heed Einstein: "Everything should be made as simple as possible, but not simpler."

1543 The Copernican System

POLAND   Nicolaus Copernicus (1473–1543)

The sun is at the center of the solar system, fixed and immobile, and planets orbit around it in perfect circles in the following order: Mercury, Venus, Earth with its moon, Mars, Jupiter, and Saturn.

The Copernican system defied the dogma that the Earth stood still at the center of the universe, and set forth a new theory of a sun-centered universe.

Not only did Copernicus place the sun at the center of the solar system, but he also gave detailed accounts of the motions of Earth, the moon, and the planets that were known at the time. He said that the Earth also revolves on its own axis, which accounts for days and nights.

Copernicus had found the truth, but to convince the world was an onerous task. He did not publish his findings because they were thought to contravene the teachings of the Catholic Church. Religious leaders of his time were against him. Martin Luther (founder of the Lutheran Church in Germany) denounced him as "a new astrologer ... the fool" who wanted "to overturn the entire science of astronomy." His book *De Revolutionibus Orbium Coelestium Libre VI* (*Six Books Concerning the Revolutions of the Heavenly Orbs*) was published at the very end of his life, and a copy placed on his deathbed. Thus the greatest astronomer of his time died without seeing his book in print— the book which ranks with Newton's *Principia* (pp. 38 and 40) and Darwin's *The Origin of Species* (p. 98) as a product of scientific genius.

See also Galileo's Concept of the Solar System, p. 28.

1577

Brahe's Theory of the Changing Heavens

Tycho Brahe (1546–1601)

The heavens are changeable, and comets move through space. The Earth is the center of the universe, and round it rotates the moon and the sun. The planets orbit the sun.

Up to now it had been believed that planets were carried on kinds of spheres that fit tightly around each other.

Brahe dissented from the Copernican doctrine (p. 19) and accepted the dogma that the Earth stood still. His real contribution to astronomy was as an observer, rather than as a theorist. He accurately measured the position of 777 stars, a remarkable achievement considering it was done without a telescope. He also measured the movement of planets, but was unable to determine their orbits. His observations paved the way for the discoveries of his assistant, Kepler

(p. 23). After Brahe's death Kepler inherited Brahe's vast accumulation of planetary observations.

Brahe's observations of the supernova, or exploding star, of 1572 and the comet of 1577 convinced him that the universe was not unchangeable as was believed by philosophers of his time. The notion of celestial spheres was not possible because comets moved through these spheres. But he still placed the Earth at the center of the universe. His contemporary, the Italian philosopher Giordano Bruno (1548–1600), believed in the sun-centered Copernican system, and for these heretical beliefs he was burned at the stake.

1600

Gilbert's Theory of Magnetism

William Gilbert (1544–1603)

A compass needle points to the magnetic poles of the Earth, which acts as a giant bar magnet. The needle would dip down at different angles at different latitudes, but would point straight down at the North Pole.

Until Gilbert proposed his theory, a magnet was a mysterious stone to scientists. Some even believed that a compass needle points to the heavens.

In 1600, Gilbert published a book, *De Magnete*, in which he described valuable facts and experiments on magnetism and static electricity. He proved that the Earth is a magnet and showed that an iron rod can be magnetized by forging it in the north-south direction. He discovered that if a magnet is cut in half, each half becomes a magnet.

Gilbert also discovered that when rubbed with a suitable substance, amber and many other substances acquire a strange power of attraction. He called such substances "electrics" (from the Greek *elektron*, for "amber"). He also invented an electrical apparatus, the forerunner of a simple electroscope, to conduct experiments on the electrical properties of various substances.

Gilbert always performed his experiments with great care, and noted down every observation he made. *De Magnete*, a great exposition of the new scientific methods, is often considered to be the first great scientific work in English. A year after its publication he was appointed court physician to Queen Elizabeth I.

Star of Bethlehem

Johannes Kepler (1571–1630)

The Star of Bethlehem was a planetary conjunction (close approach of planets in the sky) of Jupiter and Saturn in the evening sky.

Kepler was the first to identify the Christmas star with a precise event and date.

The Star of Bethlehem leading the three Magi to Jesus' birthplace is a standard symbol of Christmas. For two millennia, astronomers, theologians, believers, and sceptics have pondered the story of the star that is supposed to have announced the Christian era.

In December 1603, Kepler was intrigued by the planetary conjunction of Jupiter and Saturn. With his characteristic patience and accuracy, he began computing the planetary positions at the time of the birth of Jesus. His calculations showed that there was a triple planetary conjunction of Jupiter and Saturn in 7 BC on May 27, October 5 and December 1.

Critics of this theory say that there is a major flaw: the Bible refers specifically to a "star," not a planet or a pair of planets. Now, other theories also compete with Kepler's theory: it was (1) the planet Venus, (2) a supernova, or exploding star, or (3) a comet. Or was there really a new star—a holy light— that guided the three wise men to the manger in Bethlehem? Or was it a myth adduced by overzealous partisans?

See also Kepler's Laws of Planetary Motion, p. 23.

Kepler's Laws of Planetary Motion

Johannes Kepler (1571–1630)

First Law: The planets move in elliptical orbits with the sun at one focus. *Second Law*: The straight line joining the sun and any planet sweeps out equal areas in equal periods of time. *Third Law*: The squares of the orbital periods of the planets are proportional to the cubes of their mean distances from the sun.

Modern measurements of the orbits of the planets show that they do not precisely follow these laws; however, their development is considered a major landmark in science.

The first two laws were published in 1609 and the third in 1619. Their publication put an end to Ptolemy's cycles and epicycles (p. 16). Kepler's ardent faith in the Copernican system (p. 19)— "The sun not only stands at the center of the universe, but is its moving spirit," he asserted—brought him the disfavor of religious leaders, and the title of "mad stargazer" from the people.

Kepler was a versatile genius who, besides discovering these three laws, compiled the tables of star positions, developed the astronomical telescope, worked on infinitesimal calculus and logarithms, founded the science of geometrical optics, studied the anatomy of the human eye, explained the tides of the oceans, and wrote in Latin the first sciencefiction story, "Somnium," in which he dreamed of building a ship to sail the oceans of space in the universe.

See also Star of Bethlehem, p. 22.

1620 Bacon's Scientific Method

ENGLAND   Francis Bacon (1561–1626)

Scientific laws must be based on observations and experiments.

Bacon rejected Aristotle's deductive, or *a priori*, approach to reasoning, and suggested his inductive, or *a posteriori*, approach. Bacon discovered the most important tool of science—the scientific method—but he did not make any significant scientific discovery. "I shall content myself to awake better spirits like a bell-ringer, which is first up to call others to church," he once wrote to a friend. Bacon's bell is still ringing.

Bacon, a philosopher, advocated a new method of inquiry, completely different to the philosophical methods of ancient Greeks, in his book *Novum Organum*, which has influenced every scientist since its publication in 1620. The essence of his method is as follows: collect masses of facts by observations and experiments, analyze facts by drawing up tables of negative, affirmative, and variable instances of the phenomenon, draw hypotheses from the evidence, collect further evidence to proceed toward a more general theory. The most important aspect of this method was the idea of drawing up tentative hypotheses from available data and then verifying them by further investigations. "A true and fruitful natural philosophy has a double scale or ladder ascendant or descendant, ascending from experiments to axioms and descending from axioms to the invention of new experiments," he wrote in *Novum Organum*.

See also the appendix, under "The Scientific Method," p. 202.

1621

Snell's Law

Willebrord von Roijen Snell (1580–1626)

During refraction of light, the ratio of the sines of the angles of incidence (*i*) and refraction (*r*) is a constant equal to the refractive index of the medium.

In equation form, $n_1 \sin i = n_2 \sin r$, where n_1 and n_2 are the respective refractive indices of the two media. The refractive index of a substance is a measure of its ability to bend light. The higher the number the better light is refracted. For example, the refractive index of diamond, 2.42, is the highest of all gems.

Refraction is the change in direction of a ray of light when it crosses the boundary between the two media. It happens because light has different speeds in different media. A ray of light entering a medium where the speed of light is less (from air to water, for example) bends toward the perpendicular to the boundary of the two media (*see* diagram). It bends away from the perpendicular when it crosses from water to air. Refraction was known to ancient Greeks, but Snell, a Dutch mathematician, was the first to study it. Refraction is responsible for many optical illusions such as a stick in water appearing to be bent.

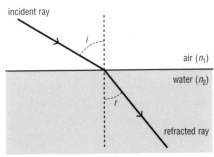

Refraction

How a ray of light bends when it passes from air to water.

1628

Harvey and the Circulation of Blood

William Harvey (1578–1657)

The heart is a pump made of muscles. It causes the blood to circulate throughout the body through the arteries and then back to the heart through veins.

We now take this for granted, but until Harvey's time people were not aware of blood circulation; they believed that blood was formed in the liver, flowed through the septum (the dividing wall) of the heart, and was then absorbed by the body.

When Harvey published this theory in his book *An Anatomical Exercise Concerning the Motion of the Heart and Blood in Animals*, it was rejected by his fellow physicians because they maintained that his ideas had not cured a single patient. John Aubrey, the gossipy biographer of the time, wrote: "I heard Harvey say that after his book came out, he fell mightily in his practice. 'Twas believed by the vulgar that he was crack-brained, and all the physicians were against him. I knew several doctors in London that would not have given threepence for one of his discoveries." It took nearly half a century before Harvey's discoveries, which were the result of painstaking experiments on animals, were accepted by the medical profession.

Harvey was court physician to Charles I, a post he held until Charles was beheaded. Aubrey again: "He was present at the Battle of Edgehill and had charge of the young Prince of Wales and Duke of York; during the battle he sat under a hedge and read a book."

1632

ITALY

Galileo's Laws of Falling Bodies

Galileo Galilei (1564–1642)

Discounting air resistance, all bodies fall with the same motion; started together, they fall together. The motion is one with constant acceleration: the body gains speed at a steady rate.

From this law we get the equations of accelerated motion: $v = gt$ and $s = \frac{1}{2}gt^2$, where v is the velocity, g is the acceleration produced by gravity and s is the distance traveled in time t.

The Greek philosopher Aristotle (384–322 BC) was the first to speculate on the motion of bodies. He said that the heavier the body, the faster it fell. It was not until 19 centuries later that this notion was challenged by Galileo. He conducted meticulous experiments on inclined planes to study the motion of falling bodies. From these experiments he formulated his laws of falling bodies.

The laws were published in his book, *Dialogue Concerning the Two Chief World Systems*, which summarized his work on motion, acceleration, and gravity.

Legend has it that Galileo gave a demonstration, dropping a light object and a heavy one from the top of the Leaning Tower of Pisa. The demonstration probably never happened, but in 1971 Apollo 15 astronauts reperformed Galileo's experiment on the moon. Astronaut David Scott dropped a feather and a hammer from the same height. Both reached the surface at the same time, proving that Galileo was right.

See also Galileo's Concept of the Solar System, p. 28.

1632

Galileo's Concept of the Solar System

Galileo Galilei (1564–1642)

The Earth and the planets not only spin on their axes; they also revolve about the sun in circular orbits. Dark "spots" on the surface of the sun appear to move; therefore, the sun must also rotate.

Galileo confirmed and advanced Copernicus' sun-centered system by observing the skies through his refracting telescope, which he constructed in 1609.

When Galileo published his masterpiece, *Dialogue Concerning the Two Chief World Systems*, which eloquently defended and extended the Copernican system (p. 19), his theories were thought to contravene the teachings of the Catholic Church. In 1633 he was tried for heresy by the Inquisition and forced to renounce his theories. In the folklore of science the trial of Galileo is as popular as Newton's apple (p. 38). Tradition has it that he was so convinced that it is the Earth that moves that the moment he was released after recanting his belief in the Copernican system, Galileo stamped his foot and muttered inaudibly "*E pur si muove*" ("And yet it moves"). Galileo was punished by confinement to his house for the last years of his life.

After Galileo's death, scientific thought gradually veered around to the idea of the sun-centered solar system. In 1992, after more than three and a half centuries, the Vatican officially reversed the verdict of Galileo's trial.

See also Galileo's Law of Falling Bodies, p. 27.

1637

FRANCE

1993

US

Fermat's Last Theorem

Pierre de Fermat (1601–65)
Andrew Wiles (b. 1953)

This theorem proves that there are no whole-number solutions of the equation $x^n + y^n = z^n$ for _n_ greater than 2.

The problem is based on Pythagoras' Theorem (p. 7): in a right-angled triangle, the square of the hypotenuse is equal to the sum of the squares on the other two sides, that is, $x^2 + y^2 = z^2$. If x and y are whole numbers, then z can also be a whole number, for example, $5^2 + 12^2 = 13^2$. If the same equation is taken to a power higher than 2, such as $x^3 + y^3 = z^3$, then z can never be a whole number.

In about 1637 Fermat, a well-known mathematician, wrote an equation in the margin of a Greek book. Then he added, "I have discovered a truly marvelous proof, which this margin is too small to contain". The problem, now called Fermat's Last Theorem, has baffled mathematicians, including the greatest, for 356 years.

In 1993 Wiles, a professor of mathematics at Princeton University, finally proved the theorem. Wiles, born in England, dreamed of proving the theorem ever since he read it at the age of ten in his local library. It took him seven years of dedicated work to prove it. The 130-page proof was published in the journal _Annals of Mathematics_ in May 1995.

Pascal's Law

Blaise Pascal (1623–62)

When pressure is applied anywhere to an enclosed fluid, it is transmitted uniformly in all directions.

The law (sometimes referred to as a principle) has practical applications in devices such as hydraulic brakes in a car, and hydraulic chairs.

The following diagram shows how a small force becomes a large force when we increase the area of one piston (the output, the force lifting the car) relative to the area of the other (the input, the force exerted by the man).

Pascal, a mathematician and philosopher, was one of the founders of the probability theory (p. 32). The unit of pressure, pascal (Pa), honors him.

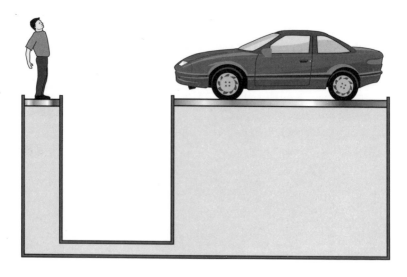

1654

Guericke's Demonstration of a Vacuum Pump

Otto von Guericke (1602–86)

The demonstration proved that air exerts pressure.

The notion that air exerts pressure may sound simple to you, but in the seventeenth century it was a remarkable discovery.

The Greek philosopher Aristotle (384–322 BC) said that "nature abhors a vacuum." This thought persisted for nearly two millennia until scientists in the seventeenth century started to study the true nature of vacuum. One of them was Guericke, a wealthy amateur scientist who in 1650 made a great technical advance by inventing a vacuum pump (sometimes called an air pump), a device to mechanically remove air from a container.

In 1654 Guericke, who was then the mayor of Magdeburg, performed a spectacular experiment before Emperor Ferdinand III and his courtiers. He used his vacuum pump to evacuate air from two brass hemispheres about 18 inches in diameter, whose edges were fitted exactly together. Eight horses were harnessed to one cup and eight to the other. But the team of 16 horses could not pull the hemispheres apart. To amaze his audience further, he turned the tap, and as the air rushed into the sphere, he effortlessly pulled the cups apart.

Guericke not only used his vacuum pump to demonstrate the power of atmospheric pressure; he also used it to show that light, but not sound, can travel through a vacuum, and that one cannot light a candle in a vacuum.

1654

Probability Theory

Blaisé Pascal (1623–62)

The study of the likelihood of an event.

Chance is something that happens in an unpredictable way. **Probability is the mathematical concept that deals with the chances of an event happening.**

The Chevalier de Méré, a seventeenth-century high-living nobleman and gambler, liked to bet even money that a six would come up in four rolls of a die. But when he started betting even money that a six would come up at least once in 24 rolls with two dice, his luck deteriorated. He asked his mathematician friend Pascal why he was having bad luck in his new game. Pascal wrote to fellow mathematician Pierre de Fermat (p. 29) about this problem and their correspondence on the matter led to the birth of probability theory.

Probability theory can help you understand everything from your chances of winning a lottery to your chances of being struck by lightning. You can find the probability of an event by simply dividing the number of ways the event can happen by the total number of possible outcomes. For example, the probability of drawing an ace from a well-shuffled pack of cards is 4/52 or 0.077 (52 cards including four aces in a pack). Below are some of the words we use to describe probability:

Description	Percentage	Probability
Absolutely certain	100%	1
Very likely	90%	0.9
Quite likely	70%	0.7
Evens (equally likely)	50%	0.5
Not likely	30%	0.3
Not very likely	20%	0.1
Never (absolutely no chance)	0%	0

1660   Hooke's Law of Elasticity

ENGLAND   Robert Hooke (1635–1703)

Within the limits of elasticity, the extension (strain) of an elastic material is proportional to the applied stretching force (stress).

The law applies to all kinds of materials, from rubber balls to steel springs. The law helps define the limits of elasticity of a material.

In equation form, the law is expressed as $F = kx$, where F is force, x change in length and k a constant. The constant is now known as **Young's modulus**, after Thomas Young (p. 62) who in 1802 gave physical meaning to k.

Hooke, a contemporary of Newton, was one of the founders in 1662 of the Royal Society in London and served as its curator of experiments until his death. Newton disliked Hooke's combative style and refused to attend Royal Society meetings while Hooke was a curator. Hooke mistrusted his contemporaries so much that when he discovered his law he published it as a Latin anagram, *ceiiinosssttu*, in his book on elasticity. Two years later, when he was sure that the law could be proved by experiments on springs, he revealed that the anagram meant *Ut tensio sic vis*, "That is, the Power of any Spring is in the same proportion with the Tension thereof … Now the Theory is very short, so the way of trying it is very easie." He was right; all you need is a spring, a ruler, and a set of weights.

1662 Boyle's Law

ENGLAND   Robert Boyle (1627–91)

The volume of a given mass of a gas at constant temperature is inversely proportional to its pressure.

In other words, if you double the pressure of a gas, you halve its volume. In equation form, $pV = $ constant, or $p_1V_1 = p_2V_2$, where the subscripts 1 and 2 refer to the values of pressure and volume at any two readings during the experiment.

Born at Lismore Castle, Ireland, Boyle was the seventh son of the first Earl of Cork. He was an extraordinarily bright child. At the age of 14 he visited Italy to study the works of Galileo. There, he decided that he would spend his life in science. In 1661 he published his most famous work, *The Sceptical Chymist*, in which he rejected Aristotle's four elements—earth, water, fire, and air— and proposed that an element was a material substance and that it could be identified only by experiment. In 1662 he made an efficient vacuum pump, now familiar to all schoolchildren, which he used to establish his law. He also used his pump to experiment on respiration and combustion and showed that air was necessary for life as well as for burning.

Boyle, who was once introduced to the famous English diarist Samuel Pepys as "son of the Earl of Cork and father of modern chymstry," established chemistry as a science—an experimental science. He believed in "setting up experiments and making observations" and not "proclaiming any theories without having tested the relevant phenomena."

See also Charles' Law, p. 54.

1668

ITALY

Redi and the Theory of Spontaneous Generation

Francesco Redi (1626–97)

From ancient times it was believed that life could arise instantaneously from living or non-living matter: maggots from decaying meat; caterpillars from leaves; frogs from slime.

Redi disproved this popular theory of spontaneous generation.

Redi prepared eight flasks, each with a variety of meat in it: a dead snake, some fish, and pieces of veal. He sealed four jars and left the other jars open to air. After a few days he found that only the open jars bred maggots. The meat in sealed jars was just as putrid but without maggots. He now repeated the experiment, but this time covering four jars with gauze tops instead of sealing them. Air could enter the jar, but not flies. Again, maggots appeared in open jars only. From these experiments Redi concluded that maggots were not formed by spontaneous generation but came from eggs laid by flies.

Even Redi's experiments could not shake people's belief in this age-old idea. The idea was finally laid to rest when in 1865 Louis Pasteur (p. 104) performed a series of experiments that showed that microorganisms are present everywhere. He established that all life comes from other life.

Redi, a physician and poet, firmly believed in the new scientific method. His simple but carefully planned experiments laid the foundations of experimental biology.

1684 Leibniz's Calculus

GERMANY Gottfried Leibniz (1646–1716)

"A new method for maxima and minima, as well as tangents ... and a curious type of calculation," that's how Leibniz introduced calculus.

Calculus has now become an important branch of mathematics dealing with the behavior of functions.

The spat between Newton and Leibniz over which of them had first devised calculus is well known to readers of science history. Newton (pp. 38–40) invented calculus (he called it fluxions) as early as 1665, but did not publish anything until 1687. The controversy continued for years, but it is now thought that each developed it independently. However, terminology and notation of calculus as we know it today is due to Leibniz. For example, Leibniz proposed the symbol ∫ (an elongated s) for summation, which is still used today. He also introduced many other mathematical symbols: the decimal point, the equal sign, the colon (:) for division and ratio, and the dot for multiplication.

Leibniz, a mathematician and philosopher, dreamed of creating a universal symbolic language which could be used for determining the truth of any proposition. "There would be no more need of dispute between two philosophers than between two accountants," he said. "It would suffice for them to take their pencils in their hands ... and say to each other ... 'Let's calculate.'" Leibniz's dream gave birth to modern mathematical logic.

1686 Ray's Concept of Species

ENGLAND   John Ray (1627–1705)

A species is a population of organisms consisting of similar individuals capable of breeding among themselves.

Ray was the first to use the word species (the Latin for "kind" or "form") in its current scientific sense.

Ray's concept of species was based on the study of over 18,600 species of plants, which he described in his three volume *Historia Plantarum Generalis*, published between 1686 and 1704. In this book he explained: "After a long and considerable investigation, no surer criteria for determining species has occurred to me than the distinguishing features that perpetuate themselves in propagation from seed. Thus, no matter what variations occur in the individual or the species, if they spring from the seed of the one and the same plant, they are accidental variations and not such as to distinguish a species." He was also the first to classify flowering plants into monocotyledons and dicotyledons (the names refer to the number of cotyledons, or seed leaves, present in their seeds), and introduced the terms "petal" and "pollen." In his other works he also wrote about fish, birds, reptiles, mammals, and fossils.

The term "species" is now used for both living and extinct organisms. Throughout the world about 1.5 million species have been identified and named. The estimates for the total number of species on the planet range from 10 to 30 million.

See also Linnean System of Classification, p. 44.

1687 Newton's Law of Gravitation

ENGLAND   Isaac Newton (1642–1727)

Any two bodies attract each other with a force proportional to the product of their masses and inversely proportional to the square of the distance between them.

The force is known as gravitation. It holds chairs to the floor and planets in their orbits.

The law may be expressed by the equation $F = GmM/r^2$, where F is force, m and M the masses of two bodies, r the distance between them, and G the gravitational constant.

While sitting in his garden, Newton noticed an apple drop from a tree. This set him to wondering why it should fall straight down to the ground. He came to the conclusion that the apple fell downward because some force was pulling it. This chance observation led him to his great theory of gravitation. (Gravity is the same thing as gravitation. The word gravity is particularly used for the attraction of the Earth for other objects.)

This is perhaps the most popular anecdote in science. The anecdote—probably apocryphal—does not tell us whether the apple hit Newton on his head or not, but it is well known that it hit upon a scientific idea that solved many mysteries of the universe. Newton published his law of gravitation in his magnum opus *Philosophiae Naturalis Principia Mathematica* (Mathematical Principles of Natural Philosophy) in 1687. The publication of *Principia*, which was written in Latin, is considered to be one of the most important events in the history of science.

The first book of *Principia* states the laws of motion (p. 40) and deals with the general principles of mechanics. The second book is concerned mainly with the motion of fluids. The third book is considered the most spectacular and explains gravitation. Newton showed that a single universal force (a) keeps the planets in their orbits around the sun, (b) holds the moons in their orbits, (c) causes objects to fall, (d) holds objects on the Earth, and (e) causes the tides.

Why do two objects attract each other? Even the great Newton could not find any explanation. "I frame no hypotheses," he said. However, Newton said that the law of gravitation is universal: that is, it applies to all bodies in the universe. The universality of the law of gravitation was challenged in 1915 when Einstein published his famous theory of general relativity (p. 140).

1687 Newton's Laws of Motion

ENGLAND   Isaac Newton (1642–1727)

First Law: An object at rest will remain at rest and an object in motion will remain in motion at that velocity until an external force acts on the object. *Second Law*: The sum of all the forces (*F*) that act on an object is equal to the mass (*m*) of the object multiplied by the acceleration, or *F = ma*. *Third Law*: To every action, there is an equal and opposite reaction.

The first law introduces the concept of inertia, the tendency of a body to resist change in its velocity. The inertia of an object is related to its mass. The second law explains the relationship between mass and acceleration. The third law shows that forces always exist in pairs.

Newton's laws of motion were also published in his book *Principia* (p. 38). These laws are so fundamental that every student of science has to learn them. When Newton was working on *Principia*, he was a professor at Cambridge University, where he was expected to read a lecture on mathematics each week. But he was so preoccupied with his work that he disregarded this obligation far more than he fulfilled it. "When he did lecture, students were scarce," writes James Gleick in *Isaac Newton* (2001). "Sometimes he read to a bare room or gave up and walked back to his chambers".

1690

Huygens' Principle

HOLLAND   Christiaan Huygens (1629–95)

Every point on a wavefront can act as a new source of waves.

Huygens originated the idea of wavefronts and used this idea to explain the propagation of waves. This principle is still used to determine the future position of a wavefront. A line perpendicular to the wavefronts is called a ray, and this ray shows the direction of the wave.

Newton (pp. 38–40) was the first to define the nature of light: light consists of tiny particles called "corpuscles" which travel in the "ether" of space. However, Huygens, a contemporary of Newton, said that light consists of waves having wavefronts perpendicular to the direction of their motion.

Huygens got his idea about waves when as a boy he watched the ripples on the canals about his home. In 1690, in his book *Traité de la Lumière* (Treatise on Light), he pointed out that light waves can pass through each other and that this property enables a person to look another person in the eye. "If light consisted of particles, the particles from one eye would collide with those from the other," he said. In his book he also developed a technique, now known as Huygens' Principle, for predicting the future position of a wavefront when its earlier position is known.

Huygens also discovered Saturn's rings (1655) and Titan, the first moon of Saturn (1665), and invented the pendulum clock (1656).

See also Young's Principle of Interference, p. 62.

1729

Biological Clock

FRANCE · Jean-Jacques d'Ortous de Mairan (1678–1771)

Some functions in plants are not regulated by the sun, but by some mechanism within the plants.

A lot more work has been done on the biological clock since de Marian's hypothesis. We now know that biological clocks are an internal timing system that regulates metabolism in all forms of life.

De Marian was an astronomer. After his experiments with plants, he went back to watching the sky. We must thank him for this digression, and also for describing in 1749 a Chinese refrigerator which used the cooling effect of evaporation. However, de Marian did not know that we are all captives of our biological clocks. Hundreds of cellular, physiological, and behavioral patterns have been observed to follow a 24-hour cycle in humans. For this reason the biological clock is also called **circadian rhythm** (from the Latin *circa diem*, about a day). Circadian rhythms are not related to the popular, pseudoscientific concept of biorhythms. Body temperature is a good example of circadian rhythms. A temperature of 98.6°F is considered to be normal body temperature, but healthy individuals show a 24-hour cycle that varies from 95.9 to 101.3°C. The body temperature is at its lowest in the early hours of morning and reaches its maximum in the late afternoon and early evening.

Jet lag and health problems associated with working in rotating shifts are caused largely by the body's battle against its circadian clock, the light-sensitive timepiece which also regulates sleep cycles. Defective clocks can trigger depression and sleep disorders. The period of circadian rhythms is rarely exactly 24 hours, but varies from 23 to 25 hours. The human internal sleep–wake cycle is about 25 hours long. Because of the 25-hour sleep–wake cycle, people are constantly advancing their sleep by an hour a day to conform to the Earth's 24-hour schedule. But when people rotate shifts, the change in sleep–wake cycle is too dramatic; the system becomes desynchronized and begins to "free run," to drift forward on its 25-hour cycle until it is back in phase. The same desynchronization is the cause of jet lag. Because the 25-hour sleep–wake cycle tends naturally to delay sleep, it is somewhat easier to adjust to work schedules that require us to stay up later than usual—a forward rotation.

The biological clock enables organisms to be in tune with their environment. Without it, their survival in a hostile environment would not be possible. Since all forms of life possess the biological clock and it is to their advantage to have the clock, it is likely that it has been developed during evolution. In vertebrates, for instance, the clock system arose more than 450 million years ago.

If we do have a biological clock, where is it located? In mammals, including humans, the clock resides in the hypothalamus of the brain in a tiny collection of cells called the suprachiasmatic nucleus (SCN). The SCN lies close to the optic tract and is directly connected to the eyes. SCNs have also been discovered in other tissues of the body.

The SCN is but one part of the so-called circadian axis; the other two components are the pineal gland (which produces the hormone melatonin in the dark) and the retina. In some people the excessive secretion of melatonin during the long dark nights (and dark) days of winter can trigger a condition known as seasonal affective disorder (SAD) or winter blues. This depressive state can be cured by exposure to suitable bright lights.

1735

Linnean System of Classification

Carl Linnaeus (1707–78)

A system for naming organisms by assigning them scientific names consisting of two parts.

This system provides a concise, orderly method for classification. The system is still used widely but now the genetic code of an organism provides a better method for classification.

In the Linnean system, also known as "binomial nomenclature," each species is given a two-word scientific name (in Latin)—the genus name, which comes first and begins with a capital letter, and the species name, which begins with a lowercase letter. Often the genus name is abbreviated, and the names are always written in italics or underlined. For example, the scientific name for humans is *Homo sapiens* (for short, *H. sapiens*). The Linnean system has six classification categories—in descending order: kingdoms, phyla, classes, orders, genera, and species—but only two are used for naming organisms.

Linnaeus became interested in botany when he was eight and gained the affectionate nickname of "the little botanist." In 1735 he published his book *Systema Naturae,* in which he introduced his classification system. "The little botanist," who became the founder of the science of taxonomy, showed his self-deprecating humor when he described *Linnaea boralis*, a plant named after him: "*Linnaea* was named by the celebrated Gronovius and is a plant of Lapland, lowly, insignificant, disregarded, flowering but for a brief space—from Linnaeus, who resembles it."

1738
SWITZERLAND

1860
SCOTLAND

Kinetic Theory of Gases

Daniel Bernoulli (1700–82)
James Clerk Maxwell (1831–79)

Gases are composed of molecules which are in constant random motion, and their properties depend upon this motion.

The theory also provides a model for the other two states of matter—liquids and solids.

Bernoulli proposed a "bombardment theory" which stated that a gas consisted of tiny particles in rapid, random motion. These moving particles would produce pressure by bombarding the container. Heating a gas would make its particles move faster. After 120 years Maxwell polished Bernoulli's ideas into a rigorous mathematical theory. In simple words, the volume of a gas is simply the space through which molecules are free to move. Collisions of the molecules with each other and the walls of a container are perfectly elastic, resulting in no decrease in kinetic energy. The average kinetic energy of a gas increases with an increase in temperature and decreases with a decrease in temperature. The theory has now been extended to liquids and solids

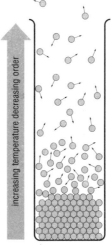

GAS
extreme disorder: molecules move at great speeds

LIQUID
some disorder: molecules free to move

SOLID
highly ordered arrangement: molecules vibrating

increasing temperature decreasing order

(see diagram).

Molecules in the three states of matter.

See also Maxwell's Equations, p. 102 and Bernoulli's Principle, p. 46.

1738

SWITZERLAND

Bernoulli's Principle

Daniel Bernoulli (1700–82)

As the velocity of a liquid or gas increases, its pressure decreases; and when the velocity decreases, its pressure increases.

The principle is expressed by a complex equation, but it can be summed up simply as the faster the flow the lower the pressure.

Bernoulli's principle has many applications. For example, it is used in the design of aircraft wings. The wing's curved upper surface is longer than the lower one, which ensures that air has to travel further and so faster over the top than it does below the wing (see diagram). Hence the air pressure underneath is greater than on the top of the wing, causing an upward force, called *lift*.

If there is a narrow constriction (throat) in a pipe or a tube, the speed of a gas or liquid is increased in the constriction, but its pressure is decreased, according to Bernoulli's principle. This effect is named the **Venturi effect** (and a pipe or tube with a narrow constriction, the **Venturi tube**) after the Italian scientist G. B. Venturi (1746–1822) who first observed it in constrictions in water channels. A perfume spray bottle or atomizer works on the same principle.

Bernoulli belonged to a highly distinguished family of scientists. His father Johann and uncle Jakob were eminent mathematicians. Five other members of the Bernoulli family were also scientists.

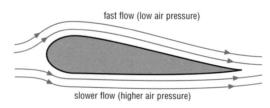

fast flow (low air pressure)

slower flow (higher air pressure)

See also Kinetic Theory of Gases, p. 45.

The Celsius Temperature Scale

Anders Celsius (1701–44)

The temperature difference between the freezing point and the boiling point of water is a hundred degrees.

The scale was called centigrade ("hundred steps") scale, but was renamed Celsius in 1969.

Celsius was an astronomer, but he is not known for his astronomical achievements, which include observations of eclipses, determining the brightness of 300 stars, and the systematic observation of the aurora borealis (beautiful displays of colored lights in the far northern skies).

His greatest accomplishment is the temperature scale he devised. In his original version, he assigned 0 to the boiling point of water and 100 to the melting point of ice. After his death, Carl Linnaeus (p. 44) reversed the scale to its present form. In 1969 the Conference Générale des Poids et Mesures, the international body responsible for the International System of Units, decreed that the centigrade scale should be called the Celsius scale.

Professor Celsius (he was appointed professor at Uppsala when he was only 29 years old) can rest in peace that an anagram for his name "Anders Celsius" is "Scale's in use, Dr" and his scale is being used worldwide. However, the **Fahrenheit scale** is still used in many countries, including the USA. In this scale, introduced in 1724 by the German–Dutch physicist Gabriel Fahrenheit (1686–1736), the freezing point of water is set at 32° and the boiling point at 212°. In scientific work the Kelvin scale (p. 92) is preferred.

The Leyden Jar

Pieter van Musschenbroek (1692–1761)
Ewald Jurgen von Kleist (1700–48)

Electricity produced by electrostatic machines can be stored in a jar.

I n modern terms, the Leyden jar is a capacitor or condenser, a device used for storing electric charge.

In 1732 Stephen Gray (c. 1666–1736), an English experimenter, discovered that electric charge could be conducted for long distances. He also classified various substances into conductors and insulators of electricity. He suggested that the best conductors were metals and thus introduced the use of electric wire. He also devised an ingenious experiment to show that the human body is a good conductor. He placed his footboy (described as "a good stout lad" by many writers of the time) face downwards on a wooden plank suspended in midair by silk cords attached to the ceiling. He charged a thick glass rod by rubbing it against a silk pillow and put it to the soles of the boy's feet. When Gray's assistant touched the boy's head with his finger, he felt a prickling sensation, which showed that the human body was a conductor.

In the mid-eighteenth century, electrostatic machines replaced glass rods for generating electricity. A typical machine consisted of a glass cylinder that rested on pivots and could be rotated by a handle. When rotated, the cylinder rubbed against a silk pillow and became charged by friction.

In 1746 Musschenbroek, a professor from Leyden in Holland, conducted a "terrifying" experiment in which he fastened one end of a brass chain to a gun barrel, the middle to an electrostatic machine, and dipped the other end in a glass jar filled with water. His assistant Andreas Cunaeus was holding the jar when Musschenbroek turned on the handle of the machine. The charge passed from machine to the gun barrel and then to the jar. Musschenbroek unknowingly touched the brass chain dipped in the glass jar with his left hand while still rotating the machine with his right hand. He was struck with such a high charge that he was paralyzed for a few moments. "For the whole kingdom of France I would not take a second shock," a terrified Musschenbroek said later. But he had made the important discovery that electricity could be stored in a jar of water. About a year earlier, von Kleist, a German scientist, also discovered the same principle independently. In later versions of the jar, which became known as the Leyden jar, water was replaced by copper foil inside and outside the jar (see diagram).

The Leyden jar became a novelty. In village fairs magicians used "electricity in a bottle" to amaze and entertain villagers. Abbé Nollet (1700–70), a French scientist, conducted a series of entertaining experiments with the Leyden jar. One day in 1746, in the presence of Louis XV and his courtiers, Nollet set up his apparatus at the Palace of Versailles and ordered 180 soldiers to join hands and form a ring with a small gap in it. The first solider in the line grasped the knob of a Leyden jar. When

the last solider in the line was ordered to touch the knob, all 180 soldiers jumped simultaneously as a charge passed through them. In another public demonstration, Nollet passed an electric charge through a 328-yard-long line of monks, each connected to the other by a short iron wire, with similar results.

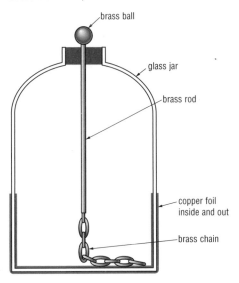

A Leyden jar.

See also Coulomb's Law, p. 52.

1772

GERMANY

Bode's Law

Johann Bode (1747–1826)

The numbers in the series 0, 3, 6, 12, 24, 48, 96, when added to 4 and divided by 10, produce the series 0.4, 0.7, 1.0, 1.6, 2.8, 5.2, 10, which gives the distances of planets from the sun in astronomical units.

The series correctly gave the distances of the six planets known at the time, except for position 2.8.

When the German–British astronomer William Herschel (1738–1822) discovered a new planet, Uranus, in 1781, it also fitted Bode's law: continuing the series by doubling 96 for Saturn, that is, 192, when added to 4 and divided by 10 gives 19.6, which is close enough to 19.2, the actual distance of Uranus from the sun in astronomical units (an astronomical unit is the mean distance from the center of the Earth to the center of the sun, about 93.2 million miles).

Bode suggested a new planet for the gap at 2.8, but in 1801 the first asteroid, Ceres, was discovered between Mars and Jupiter, at position 2.8. The planets Neptune and Pluto, discovered in 1846 and 1930, respectively, also did not fit in the positions predicted by Bode's law. Their mean distances—30 and 39.2 astronomical units—were different to the distances predicted by Bode's law, 38.8 and 77.2. (The mnemonic, My Very Educated Mother Just Served Us Nine Pizzas, may help you to remember the order of nine planets, outwards from the sun.)

1779

HOLLAND

Ingenhousz's Theory of Photosynthesis

Jan Ingenhousz (1730–99)

Green plants absorb carbon dioxide and give off oxygen, but only in the light. Plants reverse this process in the night.

The process is now known as "photosynthesis" (meaning "combining by light"). Photosynthesis enables plants to use the carbon in carbon dioxide to grow.

Photosynthesis is nature's most important chemical process: it supplies oxygen to the atmosphere and food to plants which, directly or indirectly, supply energy to most living things. Since Ingenhousz's discovery of the basic principle of photosynthesis, much has been discovered about photosynthesis. Two types of reactions take place in plants: in light reactions, solar energy is absorbed by chlorophyll, a green pigment, and converted into chemical energy, and water is split into oxygen and hydrogen. In dark reactions, carbon dioxide is converted into sugar. Thus photosynthesis converts solar energy into chemical energy needed by plants to change energy-poor compounds such as carbon dioxide and water into energy-rich sugar and oxygen.

Scientists are now looking at ways of creating artificial photosynthesis systems to harvest solar energy. Theoretically, solar energy can be tapped from these systems in two different ways: either as electrons (that is, electricity) or as hydrogen (a pollution-free fuel that can be used for heat or to make electricity). If scientists are successful in mimicking photosynthesis, we may one day have a have new source of renewable energy that does not pollute the environment.

See also Calvin Cycle in Photosynthesis, p. 180.

Coulomb's Law

Charles de Coulomb (1736–1806)

The force of attraction or repulsion between two charges is directly proportional to the product of the two charges and inversely proportional to the square of the distance between them.

The region around a charged object where it exerts a force is called its electric field. Another charged object placed in this field will have a force exerted on it. Coulomb's law is used to calculate this force.

In 1733 Charles Francois du Fay (1698–1739), a French botanist, noticed that different kinds of charges were produced by rubbing glass and rubbing gum resin: a gold leaf electrified by means of a rubbed glass tube was repelled by the glass, but attracted by rubbed gum resin. He conducted further experiments which showed that charged amber was attracted by some charged bodies but repelled by others. He concluded that there were two kinds of electricity, that he named "vitreous" (from glass) and "resinous" (from amber and resins). It was shown later that electric charges are either positive or negative and that two like charges repel each other and or two unlike charges attract each other.

Coulomb, a French physicist, made a detailed study of electrical attractions and repulsions between various charged bodies and concluded that electrical forces follow the same type of law as gravitation (p. 38). The unit of electric charge, coulomb (C), is named in his honor.

See also The Leyden Jar, p. 48.

1785

Hutton's Uniformitarian Principle

James Hutton (1726–97)

The Earth's geological phenomena can be explained in terms of natural processes such as the cycle of erosion and uplift. These processes are now at work on and within the Earth and have operated with general uniformity over an immensely long time.

The Earth has changed gradually by natural processes and will continue to change through the same processes. The discovery of uniformitarianism marked a turning point in geology; from that point geology became a science, and Hutton is now remembered as its "founder."

Hutton was trained as a physician, but he made his fortune as a farmer. While farming he became fascinated by soil, rocks, and the rolling surfaces of the Earth. He gave up farming in 1768 to pursue his interest in geology. At that time, scientists believed that the planet Earth was only a few thousand years old and only natural cataclysms could change its features. Hutton suggested that the chief agent of change was the internal heat of the Earth and proposed the uniformitarian principle, which demanded a vast expanse of time. Hutton's dictum, "we find no vestige of a beginning—no prospect of an end," caused him to be denounced as an atheist.

Hutton's original paper, presented to the Royal Society of Edinburgh in 1785, attracted little interest. In 1795 he published his ideas in a book, *Theory of the Earth*, which was also ignored by his contemporaries. Hutton's principle was revived and expanded by Lyell in 1830 (p. 75).

1787 Charles' Law

FRANCE

Jacques Charles (1746–1823)

The volume of a given mass of a gas at constant pressure is directly proportional to its absolute temperature.

In other words, if you double the temperature of a gas, you double its volume. In equation form, $V/T =$ constant, or $V_1/T_1 = V_2/T_2$ where V_1 is the volume of the gas at a temperature T_1 (in kelvin) and V_2 the new volume at a new temperature T_2.

On August 27, 1783 Charles, with the Robert brothers, made the first manned ascent in a gas balloon. The hydrogen-filled balloon ascended over Paris to a height of 3000 feet. In later flights Charles ascended to a height of 9843 feet. Charles' interest in gases led him to discover his famous law that, along with Boyle's law (p. 33), is still studied by every chemistry student.

These two laws may be expressed as a single equation, $pV/T =$ constant. If we also include Avogadro's law (p. 68), the relationship would become, $pV/nT =$ constant, where n is the number of molecules or number of "moles." The constant in this equation is called the gas constant and is shown by R. The equation—known as the **ideal gas equation**—is usually written as $pV = nRT$. Strictly, it applies to ideal gases only. An ideal gas obeys all the assumptions of the kinetic theory of gases (p. 45). There are no ideal gases in nature, but under certain conditions all real gases approach ideal behavior.

1789

Lavoisier's Law of Conservation of Mass

Antoine Lavoisier (1743–94)

In a chemical reaction, the total mass of the reacting substances is equal to the total mass of the products formed.

Mass is neither created nor destroyed in a chemical change. This law is still valid today.

Lavoisier was the first person to prove that water and air are not elements, as had been believed for centuries, but chemical compounds. He also disproved the phlogiston theory, a widely held view that when substances burn they give off "phlogiston," a weightless substance. When metals, that were considered rich in phlogiston, were burned in air they turned into a powdery substance called calx (now known as oxide). This loss of weight was explained as the loss of phlogiston. Lavoisier showed that combustion was a chemical reaction in which a fuel combined with oxygen. For his pioneering experiments Lavoisier is now remembered as the father of modern chemistry. Lavoisier's wife Marie-Anne Pierrette Paulze assisted him in much of his experimental work and illustrated his famous book, *Traité Élémentaire de Chimie* (*Elementary Treatise on Chemistry*).

Lavoisier was also involved in collecting government taxes. During the French Revolution, Lavoisier and most of the tax collectors were condemned to death. At Lavoisier's trial a request was made that the sentence might be deferred for a fortnight to enable him to finish some important experiments. "The Republic has no need for scientists, justice must take its course," the tribunal retorted. Lavoisier was guillotined.

1791 & 1799
ITALY

Galvani and Volta's Concept of Electric Current

Luigi Galvani (1737–98)
Alessandro Volta (1745–1827)

Galvani: An electric current is produced when an animal tissue comes in contact with two different metals.
Volta: An electric current is not dependent on an animal tissue and can be produced by chemicals.

Of course, Galvani was wrong and Volta was right.

Galvani, a distinguished professor of anatomy at the University of Bologna in Italy, wrote in 1792: "I am attacked by two opposition parties—the learned and the ignorant. Both laugh at me, and call me the Frog's Dancing Master. But yet I know that I have discovered one of the forces of nature." Yes, it is true that he discovered a force of nature—current electricity.

One day Galvani's wife was ill. The physician prescribed her frog soup. Galvani decided to cook it himself. Cooking experiences of husbands are worth recording. He wrote in his diary: "Sitting on my balcony, I cut up some frogs and hung their legs, that I had separated from their bodies, on an iron balustrade, before me, by means of little copper hooks which I used in my experiments. Suddenly, I saw with astonishment the frogs' legs shaking convulsively every time they chanced to touch the iron of the balcony."

It is not known how quickly Signora Galvani recovered after taking the frog soup prepared by her adoring husband, but it is well recorded in the history of science that her husband's observations paved the way for the discovery of current electricity. Galvani tried a series of experiments and finally concluded that he had discovered "animal electricity." He expounded the theory that the muscles of frogs were the source of electricity. In 1791 he published his findings in a book (_Commentary on the Effect of Electricity on Muscular Motion_).

For many years, a number of scientists performed experiments on the legs of frogs without caring for the displeasure of gourmets and thus created an artificial shortage of frogs. One of them was Volta. He could not find any electricity in the frogs' legs and therefore rejected Galvani's theory of "animal electricity." He correctly held that the metals were the source of electricity, not the frogs' legs. Volta's method of producing electric current involved using discs of copper and zinc dipped in a bowl of salt solution. He reasoned that a much larger charge could be produced by stacking several discs separated by cards soaked in salt water. By attaching wires to each end of the "pile" he successfully obtained a steady current. The "voltaic pile" was the first battery in history.

Volt, the unit of electric potential, honors Volta. And although Galvani's theory on "animal electricity" was not of any major importance, he has also achieved nominal immortality: like "volt," the words "galvanic" (sudden and dramatic), "galvanized" (iron or steel coated with zinc) and "galvanometer" (an instrument for detecting small currents) have become part of our everyday language.

See also Ohm's Law, p. 74.

1798 Rumford's Theory of Heat

ENGLAND   Benjamin Thompson (1753–1814; known as Count Rumford)

Mechanical work can be converted into heat.
Heat is the energy of motion of particles.

Other scientific developments have furthered Rumford's theory. Today, we know that heat is a form of energy associated with the random motion of atoms or molecules. Temperature is a measure of the hotness of an object.

In the eighteenth century, scientists imagined heat as a flow of a fluid substance called "caloric." Each object contained a certain amount of caloric. If caloric flowed out, the object's temperature decreased; if more caloric flowed into the object, its temperature increased. But the caloric theory failed to explain the production of heat by friction.

During the boring of a cannon, Thompson was surprised to notice that it produced a large amount of heat. He devised an experiment in which he placed a brass gun barrel and a blunt steel borer in a wooden box filled with water. Two horses turned the barrel for two and a half hours. "It would be difficult to describe the surprise and astonishment expressed in the countenances of the bystanders, on seeing so large a quantity of cold water heated, and actually made to boil, without any fire," he wrote.

American-born Rumford was probably the most colorful figure in nineteenth-century science. Among other things, he founded the Royal Institution in London and invented the calorimeter, a device for measuring heat.

1798

Malthusian Principle of Population

Thomas Malthus (1766–1834)

If unchecked, the human population would grow geometrically (1, 2, 4, 8, 16 ...), while the food supply could only grow arithmetically (1, 2, 3, 4, 5 ...). In two centuries the population would be to the food supply as 256 to 9. (In an arithmetic series of numbers there is common difference between any number and its successor, while in a geometric series each number is a constant multiple of the preceding number.)

This gloomy prediction never became a reality, mainly because of great advances in agricultural technology. Even in the twenty-first century Malthusian ideas have not been forgotten. They form the foundations of the modern theory on the relationship between economics, population, and the environment.

When Malthus published his *Essay on the Principle of Population*, it excited much public attention and placed its author, an obscure country curate, in the center of a controversial political debate on population. The essay was denounced as unholy, atheistic, and subversive of the social order.

"But science increases as fast as population ... in the most normal conditions it also grows in geometrical progression—and what is impossible for science?" remarked cofounder of communism Friedrich Engels while criticizing Malthus' essay for underestimating the element of science. Ironically, the essay had a profound influence on the progress of science. It inspired Darwin in the formulation of his theory of evolution (p. 98). In his book *The Origin of Species*, Darwin writes that his theory "is the doctrine of Malthus applied with manifold force to the whole animal and vegetable kingdoms."

1799

FRANCE

Proust's Law of Constant Composition

Joseph-Louis Proust (1754–1826)

Chemical compounds contain elements in definite proportions by mass.

In chemistry books this law is now simply referred to as the law of constant composition or the law of definite proportions.

Claude Louis Berthollet (1748–1822), then the recognized leader of science in France, rejected Proust's law. Berthollet believed that the force of chemical affinity, like gravity, must be proportional to the masses of acting substances. He suggested that the composition of chemical compounds could vary widely. Proust showed that Bertholett's experiments were not done on pure compounds, but rather on mixtures. Thus, for the first time, a clear distinction was made between mixtures and compounds.

When Dalton proposed his atomic theory (p. 65), Proust's law also confirmed the atomic theory. According to the atomic theory, atoms always combine in simple whole number ratios. For example, all water molecules are alike, consisting of two atoms of hydrogen and one atom of oxygen. Therefore, all water has the same composition. Proust's law has also been confirmed by experiments. For example, water always contains 11.2 percent hydrogen and 88.8 percent oxygen.

In recent years chemists have discovered certain rare compounds in which elements do not combine in simple whole-number ratios. These compounds are known as "berthollides." In contrast, compounds in which elements do combine in simple whole number ratios are sometimes referred to as "daltonides."

1801

Dalton's Law of Partial Pressures

John Dalton (1766–1844)

The total pressure of a mixture of gases is the sum of the partial pressures exerted by each of the gases in the mixture.

Each gas in a mixture of gases exerts a pressure, that is equal to the pressure it would exert if it were present alone in the container. This pressure is called partial pressure.

Dalton was an ardent amateur meteorologist and kept a diary—it contained 20,000 weather observations—for 57 years until the day before he died. These observations led him to the study of gases. His law of partial pressures contributed to the development of the kinetic theory of gases (p. 45).

Dalton was the son of a humble Quaker handweaver living in the tiny English village of Eaglesfield. He was educated at the village school and showed early promise of brilliance in mathematics and science. In his mid-twenties he took a position as a teacher of mathematics in Manchester, where he spent the rest of his life doing scientific research. He was the first person to do serious studies of color blindness, of which he was a victim. That's why color blindness is often referred to as Daltonism. When he died, more than 40,000 people came to pay their last respects to a modest man and a great genius.

See also Dalton's Atomic Theory, p. 65.

Young's Principle of Interference

Thomas Young (1773–1829)

Interference between waves can be constructive or destructive.

Young's principle advanced the wave theory of light of Newton and Huygens. Further advances came from Einstein and Planck.

Huygens' wave theory (p. 41) was neglected for more than a century until it was revived by Young in the opening years of the nineteenth century. Young rejected Newton's view that if light consisted of waves it would not travel in a straight line and therefore sharp shadows would not be possible. He said that if the wavelength of light was extremely small, light would not spread around corners and shadows would appear sharp. His principle of interference provided strong evidence in support of the wave theory.

He illustrated this principle with a simple experiment, which is now known to every physics students as Young's double-slit experiment. In the original experiment, Young allowed a beam of sunlight to enter a darkened room through a pinhole. The beam then passed through two closely spaced small slits in a cardboard screen. You would expect to see two bright lights on a screen placed behind the slits. But Young observed instead a series of alternate light and dark stripes. Young explained these stripes, now known as interference fringes, as the effect produced when one wave of light interferes with another wave of light. Two identical waves traveling together either reinforce each other (constructive interference) or cancel each other out (destructive interference). This effect is similar to the pattern produced when two stones are thrown into a pool of water.

Young's experiment proved conclusively that light consisted of waves, but brilliant Young—he could read at the age of two and was proficient in 14 languages by the age of 14—did not provide any mathematical explanation for his theory. That work was done by the French physicist Augustin Fresnel (1788–1827). The wave theory was further expanded by Einstein in 1905 when he showed that light is transmitted as tiny particles, or photons as they are now called, rather than waves. The current view of the nature of light is based on quantum theory (p. 128): light, as an electromagnetic radiation, is transported in photons that are guided along their path by waves. This is known as "wave–particle duality."

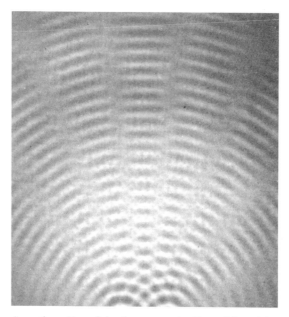

A complex pattern of simultaneous constructive and destructive interference.

1802

Howard's Classification of Clouds

Luke Howard (1772–1864)

All types of clouds can be categorized into three basic families: cirrus (hair-like), cumulus (puffs) and stratus (layers). Intermediate and compound types are: cirrocumulus, cirrostratus, cumulostratus and cumulocirrostratus or nimbus (the rain cloud).

The modern classification system is based on these basic families.

Howard, an amateur meteorologist, presented his classification (he called it "modifications of clouds") in a lecture in 1802. A detailed account was published in the *Philosophical Magazine* in 1803. Howard's theory was readily accepted by scientists of his time and he was feted as a scientific celebrity.

The first *International Cloud Atlas*, published in 1896, classified clouds by altitude but used Howard's nomenclature. In this ten-part classification the highest clouds were ranked number nine (that's why the phrase "on cloud nine," meaning extremely happy). The 1995 edition of the atlas lists the cloud terms that are currently in use as shown in the table below.

High clouds (their bases above 3.7 miles)	Middle clouds (between 1.2 and 3.7 miles)	Low clouds (below 1.2 miles)	Clouds that stretch through the three altitude bands
1. Cirrus	4. Altocumulus	7. Stratocumulus	10. Cumulonimbus
2. Cirrocumulus	5. Altostratus	8. Stratus	
3. Cirrostratus	6. Nimbostratus	9. Cumulus	

1808 Dalton's Atomic Theory

ENGLAND

John Dalton (1766–1844)

All matter is made up of atoms, that cannot be created, destroyed, or divided. Atoms of one element are identical but different from those of other elements. All chemical change is the result of combination or separation of atoms.

Of course, we now know that atoms are not indestructible or indivisible.

Now, we take atoms for granted, but when Dalton published his atomic theory, many contemporary scientists sneered at the idea of atoms. Humphry Davy (1778–1829), England's most celebrated chemist, considered the atomic theory "a tissue of absurdities." The French chemist Claude Berthollet (1748–1822) was simply "sceptical." Some even declared that he was suffering from hallucination—imagining little balls of atoms. But within a few years scientific thought came around to the idea of atoms. We now know that Dalton's theory provides one of the pillars on which the science of chemistry rests.

Dalton examined various substances in which two elements form more than one type of compound and concluded that if two elements A and B combine to form more than one compound, the different masses of A that combine with a fixed mass of B are in a simple whole-number ratio. This is known as the **law of multiple proportions**.

See also Democritus' Atomic Theory, p. 9; Dalton's Law of Partial Pressures, p. 61.

1808

FRANCE

Gay-Lussac's Law of Combining Volumes

Joseph Louis Gay-Lussac (1778–1850)

Volumes of gases that combine or that are produced in chemical reactions are always in the ratio of small whole numbers.

For example, one volume of nitrogen and three volumes of hydrogen produce two volumes of ammonia. These volumes are in the whole number ratio of 1:3:2.

Gay-Lussac was a great experimentalist and his law was based on extensive experiments, but he never bothered to explain why gases behave in the way suggested by his law. The explanation, however, came from Avogadro (p. 68).

In 1805 Gay-Lussac worked with the German explorer Alexander von Humboldt (1769–1859) on a series of experiments to find out the ratio in which hydrogen and oxygen combine to form water. For some of the experiments they required special thin-walled glass vessels, which had to be imported from Germany. At that time the French custom duty on imports was very high, and to avoid it the ever ingenious Humboldt came up with an idea: he instructed the German glassblowers to seal the long necks of the vessels and label them "German Air: Handle with Care." When the vessels arrived in France, the puzzled custom officers could not find "German Air" in their list of dutiable goods and so let the vessels through. Perhaps in a little way Humboldt's ingenuity helped formulate the law all chemistry students have to cram.

Lamarck's Theory

Jean-Baptiste Lamarck (1744–1829)

Characteristics acquired by one generation can be inherited by the next.

Lamarck's theory was rejected after the publication of the theories of Darwin (p. 98) and Mendel (p. 103).

Lamarck's theory is inextricably linked with the giraffe's neck. Textbook writers call upon the giraffe's long neck to explain Lamarck's theory: ancestral giraffes had shorter necks which they frequently stretched to reach the foliage of large trees. The slightly longer neck acquired in such a way was passed on to their offspring. Eventually the continuing inheritance of slightly longer necks gave rise to modern long-necked giraffes.

In his book *Philosophie Zoologique*, published in 1809, the year of Darwin's birth, Lamarck gave numerous examples to illustrate his theory: how the heron acquired its long legs, the giraffe its long neck, the anteater the long tongue, and so on. "It is interesting to observe the result of habit in the peculiar shape and size of the giraffe," he writes. "This animal, the largest of the mammals ... is obliged to browse on the leaves of trees and to make constant efforts to reach them. From this habit long maintained in all its race ... its neck is lengthened to such a degree that the giraffe, without standing up on its hind legs, attains a height of twenty feet." His fellow biologists, however, rejected his ideas "as worthless, unscientific speculation."

1811

Avogadro's Law

Amedeo Avogadro (1776–1856)

Equal volumes of all gases at the same temperature and pressure contain the same number of molecules.

Although Avogadro's law was correct, it remained unnoticed by his peers for nearly 50 years.

In 1811, when Avogadro proposed his law, very little was known about atoms and molecules. Avogadro realized that Gay-Lussac's law (p. 66) provided a way of proving that an atom and a molecule were not the same. He suggested that the particles (now known as molecules) of which nitrogen gas is composed consist of two atoms; thus the molecule of nitrogen is N_2. Similarly, the molecule of hydrogen is H_2. When one volume (one molecule) of nitrogen combines with three volumes (three molecules) of hydrogen, two volumes (two molecules) of ammonia, NH_3, are produced. However, the idea of a molecule consisting of two or more atoms bound together was not understood by the chemists of that time. Avogadro's law was forgotten until 1858, when the Italian chemist Stanislao Cannizzaro (1826–1910) explained the necessity of distinguishing between atoms and molecules.

From Avogadro's law it can be deduced that the same number of molecules of all gases at the same temperature and pressure should have the same volume. This number has now been determined experimentally: its value is 6.0221367×10^{23}, and we call it **Avogadro's number** (or Avogadro's constant). Every chemistry student knows this cornerstone of chemistry.

Oersted's Theory of Electromagnetism

Hans Christian Oersted (1777–1851)

Electric current produces a magnetic field.

A compass needle deflects when it is placed near a current-carrying wire.

Oersted, who was professor of physics at the University of Copenhagen, had long suspected that there was a connection between electricity and magnetism. During one of his lectures he placed a current-carrying wire at right angles to, and directly over, a compass needle. No effect was observed. After the lecture, when several students came up to meet him at the demonstration table, he accidentally placed the wire parallel to the compass needle. To his surprise, he observed the needle swing away from north and eventually come to rest perpendicular to the wire. When he switched the current off, the needle pointed north again. He reversed the current in the wire and this time the needle deflected in the opposite direction. Oersted was stunned to see this effect of electric current on magnetism.

The entire practical use of electricity in our time is founded on the connection between electricity and magnetism. And it was an "accident" that led to this truly momentous discovery. The French mathematician Joseph Lagrange (1736–1813) once observed that "such accidents come only to those who deserve them." Although Oersted had discovered electromagnetism, he did little more about it. This task was left to Ampère (p. 73).

See also Fleming's Rules, p. 118.

1823

GERMANY

Olbers' Paradox

Heinrich Wilhelm Olbers (1758–1840)

Why is the sky dark at night?

This deceptively simple question has puzzled astronomers for centuries. And no, the answer is not "because at night the sun is on the other side of the Earth."

Olbers pointed out that if there were an infinite number of stars evenly distributed in space, the night sky should be uniformly bright—with a surface brightness like the sun's. He believed that the darkness of the night sky was due to the absorption of light by interstellar space.

But he was wrong. Olbers' question remained a paradox until 1929 when astronomers discovered that the galaxies are moving away from us and the universe is expanding. The distant galaxies are moving away from us at a speed so high that it diminishes the intensity of light we receive from them. In addition, this light shifts slightly toward the red end of the spectrum. Red light has less energy than blue light. These two effects significantly reduce the light we receive from distant galaxies, leaving only the nearby stars, that we see as points of light in a darkened sky. However, in an infinite and stationary universe, uniformly filled with stars, our line of sight would always end at the surface of a star, and the entire sky should therefore be bright.

See also Hubble's Law, p. 153

1824

FRANCE

Carnot Cycle

Nicolas Sadi Carnot (1796–1832)

The Carnot Cycle is the most efficient cycle for operating a reversible heat engine. It illustrates the principle that the efficiency of a heat engine depends on the temperature range through which it works. The cycle has a four-stage reversible sequence: adiabatic compression and isothermal expansion at high temperature; adiabatic expansion and isothermal compression at low temperature. (Adiabatic = no heat flows into or out of a system; isothermal = at a constant temperature.)

The cycle played an important role in the development of thermodynamics.

Carnot was 24 when he began his studies of steam engine efficiency. Four years later he published his findings in his book *Réflexions sur la Puissance Motrice du Feu* (*Reflections on the Motive Power of Fire*), which founded the science of thermodynamics. Carnot suggested that the *puissance motrice* ("motive power"; by which he meant work or energy) of a heat engine was derived from the fall of heat from a higher to a lower temperature. Thus a heat engine was similar to a watermill that derived its motive power from the fall of water from a higher level to a lower level.

Scientists realized the full importance of *Réflexions* in 1878, when Carnot's brother discovered notes showing how his brother had provided further proof in support of his theory. The notes, written at some time between the publication of *Réflexions* and Carnot's death, also included a remarkably accurate value for the mechanical equivalent of heat (p. 90).

1827

SCOTLAND

Brownian Motion

Robert Brown (1773–1858)

Tiny solid particles suspended in a fluid are in continuous random motion.

This motion is caused by constant collisions between the suspended particles and the fluid molecules. The motion is visible in dust particles dancing in a sunbeam.

One summer day Brown was using a microscope to view very tiny pollen grains in water. He noticed that the grains moved about irregularly even though the water appeared to be perfectly still. He studied other dust particles suspended in water and noticed that they also moved at random. Brown could not explain the cause of this motion.

In 1905 Einstein studied Brownian motion mathematically and used it to calculate the approximate size and mass of atoms and molecules. We now know that the motion of the pollen grains observed by Brown was caused by random collisions of rapidly moving molecules of water.

Brown was a botanist on Matthew Flinders' voyage to Australia aboard the *Investigator* (1801–03). He returned to England with more than 4000 specimens of plants, more than half of which were unknown to science. Brown is also remembered for another discovery: the recognition of a small body within cells as a regular feature of cells. He named this the "nucleus" (from the Latin for "little nut"). Plant cells were discovered by Hooke (p. 34), who named them so because they reminded him of tiny monks' rooms in monasteries.

1827

Ampère's Law

André-Marie Ampère (1775–1836)

Two current-carrying wires attract each other if their currents are in the same direction, but repel if their currents are opposite. The force of attraction or repulsion is directly proportional to the strength of the current and inversely proportional to the square of the distance between them.

This law helped to develop a new area of study called electromagnetism.

When the French physicist François Arago (1786–1853) heard of Oersted's experiments on an electric wire and a compass needle (p. 69), he repeated them at the French Academy of Sciences. Ampère also witnessed the demonstration. Ampère generalized that the magnetic effect was the result of the circular motion of an electric current. The effect was increased when the wires were coiled. When a bar of soft iron was placed in the coil it became a magnet.

Ampère was in the habit of carrying a chalk stub in his pocket and using it on any convenient surface as a blackboard whenever a new idea came to his mind. Once the only surface he could find was the back of a hansom cab. After he had covered it with equations, he was astonished to see his "blackboard" running away from him. Ampère's name is commemorated in the unit of electric current and has become a household word (who hasn't heard of "amps"?).

See also Ohm's Law, p. 74.

1827

GERMANY

Ohm's Law

Georg Simon Ohm (1789–1854)

The electric current in a conductor is proportional to the potential difference.

In equation form, $V = IR$, where I is the current, V the potential difference, and R a constant called resistance.

When Ohm published his law, his book was called "a web of naked fancies" and the German education minister said that "a physicist who professed such heresies was unworthy of teaching science" and sacked him. Ohm has now been honored by having the unit of electrical resistance named after him. If we use units of V, I, and R, Ohm's law can be written in units as volts = ampere x ohm. It is a coincidence that this law relates three scientists of three different nationalities—Italian, French, and German.

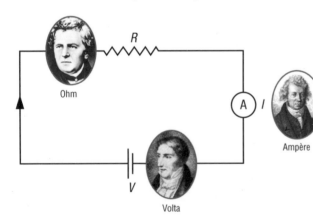

Ohm

A I

Ampère

R

V

Volta

How Ohm's law links potential difference (voltage), current, and resistance and the scientists Volta, Ampère, and Ohm.

See also Galvani and Volta's Concept of Electric Current, p. 56; Ampère's Law, p. 73.

Lyell's Theory of Uniformitarianism

Charles Lyell (1797–1875)

Considering indefinitely long periods of time, geological processes that are responsible for geological changes have always been the same as they are now.

Lyell's theory revived and expanded Hutton's uniformitarian principle (p. 53).

We now know that the Earth is about 4.6 billion years old, but in the early nineteenth century most people believed that the Earth was only about 6000 years old. They also believed that geological phenomena such as the formation of mountains and valleys were the result of cataclysmic events such as Noah's flood. This theory of Earth's history was known as "catastrophism."

In his book *Principles of Geology*, published in three volumes from 1830 to 1833, Lyell rejected these centuries-old ideas and suggested that the Earth is very old and is perpetually changing. Lyell, born in the year Hutton died, accepted Hutton's ideas on uniformitarianism and supported them with massive amounts of observational data. Darwin was so influenced by Lyell's method of supporting a theory by observational evidence that he used the same technique in his book, *The Origin of Species* (p. 98). Lyell's book is considered the most influential book on geology—no wonder its author is called the "father of modern geology." Lyell is also remembered for giving us the famous saying: "The present is the key to the past."

1831 Faraday's Law of Induction

ENGLAND Michael Faraday (1791–1867)

A changing magnetic field around a conductor produces an electric current in the conductor. The size of the voltage is proportional to the rate of change of the magnetic field.

This phenomenon is called "electro-magnetic induction" and the current produced "induced current." Induction is the basis of the electric generator and motor.

For his epoch-making experiment Faraday wrapped two coils of insulated wire around opposite sides of an iron ring. One coil was connected to a battery, the other to a wire under which lay a magnetic compass needle. He observed no effect when the current was steady, but when he turned the current on and off he noticed the needle moving. On August 29, 1831 Faraday's crude apparatus made history when it "produced electricity from magnetism."

Legend has it that William Gladstone, England's Chancellor of the Exchequer, after witnessing Faraday's demonstration of electromagnetic induction asked: "But what good is it?" Faraday famously replied: "I don't know, but one day, sir, you will be able to tax it."

The Russian physicist Heinrich Lenz (1804–65) extended Faraday's law when in 1833 he suggested the following: the changing magnetic field surrounding a conductor gives rise to an electric current whose own magnetic field tends to oppose it. This is now known as **Lenz's law**. This law is in fact Le Châtelier's Principle (p. 114) when applied to the interactions of currents and magnetic fields.

1831 Graham's Law of Diffusion

SCOTLAND   Thomas Graham (1805–69)

Under the same conditions, the rate of diffusion of a gas is inversely proportional to the square root of its density.

For example, hydrogen diffuses four times as fast as oxygen under the same conditions of temperature and pressure.

Gases have no fixed volume; they expand to fill the entire volume of their container. For example, when a perfume bottle is opened in a room a small amount of perfume evaporates and soon the perfume particles become spread more or less evenly throughout the entire room. This spreading of gas particles is called diffusion. Graham once observed that a cracked bottle inverted over water and containing hydrogen lets out the hydrogen faster than air enters. This observation led him to the conclusion that the lightest gas diffuses most rapidly. From further experiments, Graham deduced his law, now known to every chemistry student.

Graham's forceful father wanted him to enter the church, but he resisted and entered Glasgow University to study science. Graham, a brilliant experimentalist but an awful teacher, is now referred to as the father of colloid chemistry (a colloid is a kind of mixture; for example, smoke is a colloid of carbon particles suspended in air). In 1861 he invented the process of dialysis, which is based on the principle that some material will diffuse across a semipermeable membrane and some material will not. He tested the process on an ox bladder membrane that served as a kind of molecular strainer. Using this process he was able to extract urea from urine.

1832

Gauss' Law

Carl Friedrich Gauss (1777–1855)

The electrical flux through a closed surface is proportional to the sum of the electric charges within the surface.

An electric field can be pictured by drawing lines of force. The field is stronger where these lines crowd together, weaker where they are far apart. Electrical flux is a measure of the number of electric field lines passing through an area.

Gauss' law describes the relationship between electric charge and electric field. It is an elegant restatement of Coulomb's law (p. 52).

Gauss ranks with Archimedes and Newton as one of the greatest mathematicians of all time. It has been said of him that almost everything which the mathematics of the nineteenth century has brought forth in the way of original scientific ideas is concerned with the name of Gauss. "He lives everywhere in mathematics," says E. T. Bell in his authoritative biographies of mathematicians *Men of Mathematics* (1937).

Gauss was a wunderkind—a Mozart of mathematics. One Saturday in 1779, when not yet three, he watched his father making out the weekly payroll for the laborers under his charge. His father made a slip in his long calculations, and he was astonished to hear the little boy say, "Father, the calculation is wrong, it should be …" A check of the figures told his father that his "wonder-child" was right.

Galois' Theory

Évariste Galois (1811–32)

The study of solutions of some equations and how different solutions are related to each other.

This brilliant and complex theory has many applications; for example, it can be used to solve classical mathematical problems such as "Which regular polygons can be constructed by ruler and compass?" The theory has equally melodramatic origins: most of the theory was feverishly scribbled by a 20-year-old mathematician the night before his death.

At school Galois was only interested in exploring books about mathematics. At 17 he wrote his first mathematical paper and sent it to the Académie des Sciences through Augustin Cauchy, a famous mathematician, but it was lost. A year later he sent another paper to Joseph Fourier, the Secretary of the Academy, for consideration for the Academy's mathematical prize. Fourier died before he could forward the paper.

Galois was a staunch republican. During a dinner in 1831 he raised his glass with an open dagger in his hand, and was arrested for making a threat against the king. After his release from prison he fell in love with a girl, but was challenged by her fiancé to a duel. The night before the duel Galois scribbled down his theory frantically. Early next morning he received a pistol shot in the abdomen and died the next day. His last words to his brother were: "Don't cry—I need all my courage to die at twenty."

1833

Beaumont's Experiments on the Gastric Juice

William Beaumont (1785–1853)

The gastric juice is a chemical agent, and hydrochloric acid its most important component. Digestion is a chemical process.

Beaumont demolished the myth that the gastric juice was as "inert as water" and said that it was the most general solvent in nature; even the hardest bone could not withstand its action.

In 1822, when Beaumont, a US Army surgeon in Michigan, attended a patient who had been shot accidentally at close range, he found a portion of the patient's lung and stomach protruding through the external wound. He cleaned and dressed the wound but did not expect his patient to live. The 19-year-old French–Canadian Alexis St Martin was a tough young man and pulled through, but a hole remained in his stomach. Beaumont covered it with a metal plate and St Martin led an active life, married, and fathered four children.

In 1825 Beaumont realized that he had a walking laboratory in St Martin. Over the next eight years he performed hundreds of experiments (such as suspending by a silk thread a piece of food in St Martin's stomach hole and then observing its digestion at regular intervals) and gathered a vast amount of information about the gastric juice and how the stomach digests various foods. In 1833 he published his findings, *Experiments and Observations on the Gastric Juice and the Physiology of Digestion*, which made him famous throughout the medical world.

1834 Babbage's Analytical Engine

ENGLAND
Charles Babbage (1791–1871)
Augusta Ada King, Countess of Lovelace (1815–52)

A machine which, like the modern computer, had a separate store for holding numbers (memory) and a "mill" for working on them (arithmetic unit), and a punched card system for specifying the sequence of instructions (input) and for obtaining results (output).

The machine was designed for mathematical calculations only.

Babbage, a mathematician and inventor, designed three difference engines—mechanical devices that would compute and print mathematical and navigational tables—but never built one. He also designed an analytical engine. Although he prepared detailed drawings for thousands of parts, only a few parts were built. His project was far ahead of its time and the Victorian technology could not provide precise machined parts. In 1991, on the two-hundredth anniversary of Babbage's birth, the British Science Museum built the difference engine no. 2 (designed between 1847 and 1849).

The calculating section of the engine weighs 2.8 tons and consists of 2400 parts.

"It has no pretensions whatever to originate anything. But it can do whatever we know how to order it to perform," wrote Lovelace about the analytical engine. Lovelace, a mathematician, is acknowledged by many as the world's first computer programmer. Daughter of the poet Lord Byron, Lovelace worked closely with Babbage in writing instructions and publicizing his difference and analytical engines. Her writings provide the first description of programming techniques. She died of cancer at the young age of 36.

1834

Faraday's Laws of Electrolysis

Michael Faraday (1791–1867)

First Law: The amount of a substance liberated during electrolysis is proportional to the quantity of electricity passed. *Second Law*: The relative amounts of substances liberated are proportional to their equivalent weights.

The above statements are close to Faraday's original statements. Now the term "equivalent weight" has been replaced by "mole."

Faraday was the first to see the difference between substances that conduct an electric current in solution and those that do not. He coined the term "electrolyte" (loosening by electricity) and "ion" (wanderer). Though Faraday's most influential discoveries were in electromagnetism (p. 76) on which modern electric motors, dynamos, and transformers are based, his law of electrolysis started the modern electroplating industry.

Faraday was a great experimenter, in quality as well as quantity. He kept a meticulous day-to-day record of all his experiments. His notebooks reveal that in all he made 16,041 experimental observations over a period of 42 years. He used to say that he was content if one in a thousand of his experiments led to a really important discovery. He worked with his own hands, wearing an apron full of holes. His laboratory assistant was an old ex-soldier, Sergeant Anderson, whose chief virtue was blind obedience. One evening Faraday forgot to dismiss Anderson, and found him still at work next morning.

The unit of capacitance, farad (F), is named in honor of Faraday.

1835

Coriolis Effect

Gaspard de Coriolis (1792–1843)

On a rotating surface an imaginary force appears to act at right angles to the rotating Earth, causing a body to follow a curved path opposite to the direction of the Earth's rotation.

N o actual force is involved; the imaginary force is an effect of the Earth's rotation.

In the southern hemisphere, the Earth beneath us is rotating to the right all the time. Anything moving on the surface will appear to deflect to the left. In the northern hemisphere things are opposite: objects deflect to the right.

This deflection is not real: the bullet flies in a straight line while the target moves to the right. The Coriolis effect acts on all objects moving free of the Earth: the faster the object, the greater the deflection. The Coriolis effect is zero at the equator.

The Coriolis effect is significant in the atmosphere and the oceans. For example, in the southern hemisphere,

Due to Earth's rotation

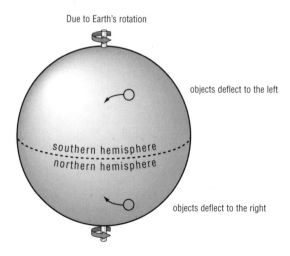

objects deflect to the left

southern hemisphere
northern hemisphere

objects deflect to the right

air moving from high to low pressure spins clockwise; while in the northern hemisphere air moving from high to low pressure spins anticlockwise. The amount of spin depends on the latitude and the speed at which the air is moving. Similarly, winds closer to the poles will be deflected more than winds at the same speed closer to the equator. However, the story that water draining out of a bath or basin rotates in the clockwise direction in the southern hemisphere, anticlockwise in the northern, is an urban myth. In the case of a draining bath or basin the Coriolis effect is negligible and is overwhelmed by other forces, such as gravity; the Coriolis force is about 30 million times weaker than gravity. Similarly, pigs' tales do not curl one way in the southern hemisphere and the reverse in the northern.

Buys Ballot's law can be used to work out the wind direction at different locations on a weather chart. Proposed in 1857 by the Dutch meteorologist Christoph Buys Ballot (1817–90), the law states that for an observer in the southern hemisphere standing with his or her back to the wind, the atmospheric pressure will be lower to his or her right than to his or her left. In the northern hemisphere, the lower atmospheric pressure will be on the observer's left. In 1853 a similar law was proposed by the American educator and meteorologist James Henry Coffin (1806–73).

Agassiz's Theory of Ice Ages

Jean Louis Agassiz (1807–73)

During the past 600 million years there have been 17 known ice ages on the Earth.

Ice ages (also known as glacial periods) are periods in history when enormous ice sheets covered large areas of land not usually covered by seasonal ice.

Holidaying in the Swiss Alps in 1836, Agassiz, a keen naturalist, noticed rocks marred by deep gouges and scrapes. He formed the novel idea that the scrapings were caused by moving glaciers and some time in the ancient past all Europe had been covered by glaciers. This led him to his theory that glaciers were the result of the ice ages. He presented his new theory to the Helvetian Society in 1837.

Much more is now known about ice ages, but scientists still do not know what causes an ice age. The list of suspects includes the Earth's orbit, drifting continents, changes in atmospheric carbon dioxide, and even cosmic rays.

The last ice age occurred in the Pleistocene epoch (that began about 2 million years ago) and was made up of four glacial periods, the last of which extended from about 40,000 to 10,000 years ago. Today we live in a relatively warmer interglacial period, as we are heading into a glacial period. In present times about 10 percent of the Earth's surface is covered by ice, but during the last ice age about 30 percent of the land was covered by ice.

1840

SWEDEN

Berzelius' Concept of Allotropes

Jöns Jacob Berzelius (1779–1848)

An element can exist in two or more forms with different properties.

The various forms are known as allotropes. For example, **graphite, diamond, and buckyballs (p. 192) are three crystalline allotropes of carbon; glass and quartz are noncrystalline and crystalline allotropes of glass.**

Berzelius converted charcoal into graphite by chemical processes and declared that some elements might exist in two or more forms with different properties. However, the Swede contributed more than just allotropes to chemistry: he gave it a new language.

When Dalton revived the idea of the atom as a unit of matter (p. 65), he used circular symbols to represent atoms. Berzelius discarded Dalton's cumbersome system and in its place introduced a rational system of chemical shorthand. "It is easier to write an abbreviated word than to draw a figure that has little analogy with words and that, to be legible, must be made of larger size than ordinary writing," he declared. "I shall therefore take as the chemical sign, the initial letter of the Latin name of each element. If the first two letters be common to two elements I shall use both the initial letter and the first letter they have not in common". Berzelius' method is now followed all over the world. And his discovery of allotropes of carbon has led chemists to discover not only more allotropes of carbon, but also allotropes of many other elements.

1842

AUSTRIA

Doppler Effect

Christian Johann Doppler (1803–53)

Any source of sound or light moving away from an observer changes in frequency with reference to the observer.

For example, the pitch of the whistle of a train changes as it runs past a person standing on a platform: it is higher when the train is approaching the person; and lower when it is moving away from the person.

Doppler explained the effect by pointing out that when the source of sound is moving towards the observer, sound waves reach the ear at shorter intervals, hence the higher pitch. When the source is moving away, the waves reach the ear at longer intervals, hence the lower pitch. The Doppler effect also occurs when the source of sound is stationary and the observer is moving.

In 1845 the Dutch meteorologist Christoph Buys Ballot (p. 84) tested the Doppler effect for sound waves in an endearing experiment—as a moving source of sound he used an orchestra of trumpeters standing in an open car of a railroad train, whizzing through the Dutch countryside near Utrecht. The experiment, of course, demonstrated the correctness of Doppler's principle.

Doppler also predicted that a similar effect would apply to light waves, but provided no explanation. In 1849 Fizeau (p. 93) showed that the Doppler effect applies to light coming from distant stars.

See also Olbers' Paradox, p. 70.

First Law of Thermodynamics

Julius Robert von Mayer (1814–78)

Heat is a form of energy and energy is conserved.

In equation form, $DU = Q - W$, where DU is the change in the internal energy of a system, Q is heat energy received by the system, and W is work done by the system (in physics, capital Greek letter delta D represents "change in" a quantity). In "thermodynamics," "thermo" refers to heat and "dynamics" to work.

The first law is one the great laws of physics. It's simply a restatement of the law of conservation of energy: energy is neither created nor destroyed, but may be changed from one form to another.

Mayer was a physician and had no background in physics. While he was working as a physician aboard a Dutch ship in the East Indies, he noticed that the blood of sailors was unusually red. He thought that the heat of the tropics increased the metabolic rate, resulting in an increase in oxygen in sailors' blood. Surplus oxygen caused extra redness. He went a step further in his reasoning: muscular exertion (work) also produced heat and there must be a relationship between work and heat.

We do not know whether Homer Simpson (of TV's *The Simpsons*) really understands the laws of thermodynamics (see also p. 95), but he has been seen storming around his house and shouting "In this house we follow the laws of thermodynamics."

1843

GERMANY

Sunspot Cycle

Samuel Heinrich Schwabe (1789–1875)

The number of visible sunspots varies in a regular cycle that averages about 11 years.

Although Galileo (p. 28) was the first to study sunspots, through his telescope in 1612, Schwabe, an amateur astronomer, made careful records of sunspots almost daily for 17 years before announcing his theory. He continued his observations for another 25 years.

Sunspots are like freckles on the sun's bright face. Wherever magnetic fields emerge from the sun, they suppress the flow of surrounding hot gases, creating relatively cool regions that appear as dark patches in the sun's shallow outer layer, known as the photosphere. Sunspots vary in size from 620 to 25,000 miles across and may last from a few days to many months.

The sunspot cycle causes "solar minimum" or "solar maximum." Near a solar minimum there are only a few sunspots. During a solar maximum there is a marked increase in the number of sunspots and solar flares, that are huge bursts of energy released from the region of sunspots. Solar flares can produce dramatic changes in the emission of ultraviolet rays and x-rays from the sun.

Ever since the cycle was documented by Schwabe, scientists have been fascinated by the possibility that the cycle might influence the Earth's weather and climate. However, this link has not yet been proven to full satisfaction.

1843

Joule's Mechanical Equivalent of Heat

James Prescott Joule (1818–89)

A given amount of work produces a specific amount of heat.

In modern terms, 4.18 joules of work is equivalent to one calorie of heat.

In 1798 Count Rumford suggested that mechanical work can be converted into heat (p. 58). This idea was pursued by Joule, who conducted thousand of experiments to determine how much heat could be obtained from a given amount of work. In one experiment he used weights on a pulley to turn a paddlewheel immersed in water. The friction between the water and the paddle wheel caused the temperature of the water to rise slightly. The amount of work could be measured from the weights, and the distance they fell gave the heat produced by the rise in temperature. Joule was so keen on his experiments that on his honeymoon in Switzerland he carried a long and sensitive thermometer to measure the temperature of the Alpine waterfalls.

Joule was the son of a brewer and started working in his father's brewery when he was 15. He was self-educated but proficient in mathematics. He learned the art of precise measurements in the brewery. All his experiments on the mechanical equivalent of heat depended upon his ability to measure extremely slight increases in temperature. Undoubtedly, he was an experimenter of the highest caliber, and we have kept his name alive today by naming the unit of work and energy, joule (J).

1845

GERMANY

Kirchhoff's Laws

Gustav Kirchhoff (1824–87)

First Law (junction law): At any junction point in an electrical circuit, the sum of all currents entering the junction must equal the sum of all currents leaving the junction.

Second Law (loop law): For any closed loop in an electrical circuit, the sum of the voltages must add up to zero.

In equation form, the first law is $I = I_1 + I_2 + I_3 + \ldots$, where I is the total current and I_1, I_2, I_3 and so on are the separate currents. Second law: $V = V_1 + V_2 + V_3 + \ldots$, where V is the total voltage and V_1, V_2, V_3 and so on are the separate voltages.

These laws are extensions of Ohm's Law (p. 74) and are used for calculating current and voltage in a network of circuits. Kirchhoff formulated these laws when he was a student at the University of Königsberg.

Kirchhoff also showed that objects that are good emitters of heat are also good absorbers. This is known as **Kirchhoff's law of radiation**. Black clothes, for example, are good absorbers of heat as well as good emitters, but wearing black clothes on a hot day would still make you hot. Why? The outside temperature is higher than the body temperature and the body receives more heat than it can emit. White clothes, on the other hand, are good reflectors and therefore poor absorbers of heat and therefore suitable for wearing on hot days.

See also Kirchhoff–Bunsen Spectroscopy Theory, p. 100.

1848

SCOTLAND

Absolute Zero

William Thomson, known as Lord Kelvin (1824–1907)

Molecular motion (or heat) approaches zero at temperatures approaching –273.15°C (–459.67°F).

This temperature is known as absolute zero. It is the theoretical lowest limit of temperature.

Like the speed of light, absolute zero can be approached closely but cannot actually be reached, as to reach it an infinite amount of energy is required. The temperature scale based on absolute zero is known as the **Kelvin scale** (Kelvin, symbol K without the degree sign). One kelvin degree equals one Celsius degree (p. 47).

The energy of a body at absolute zero is called "zero-point energy." According to Heisenberg's uncertainty principles (p. 148), atoms and molecules can exist only in certain energy levels: the lowest energy level is called the ground state and all higher levels are called excited states. At absolute zero all particles are in the ground state.

Thomson was the greatest physicist of his time. For 53 years he was professor at the University of Glasgow, but he was a failure as a lecturer and teacher. He was so preoccupied with his work that if any new idea came to his mind while lecturing, he would digress and forget all about the topic of his lecture. No, he was not the archetypical absentminded professor; he had an extraordinarily clear mind and a powerful personality. He once said: "Science is bound by the everlasting laws of honor to face fearlessly every problem that can be presented to it."

1849

FRANCE

Fizeau's Experiment on the Speed of Light

Armand Hippolyte Louis Fizeau (1819–96)

The first successful experiment to determine the speed of light.

Prior to this experiment, it was believed that light had an infinite speed.

Fizeau carried out his experiment in Paris between the belvedere of a house at Montmartre and a hill at Suresnes—a distance of 5.4 miles. He placed a rotating toothed wheel with 720 gaps at Montmartre and a mirror at Suresnes. When the wheel was at rest light passed through one gap and was reflected.

When the wheel was rotated slowly the light was completely eclipsed from the observer. When the wheel was turned rapidly the reflected light passed through the next gap. Fizeau observed this at a maximum speed of 25 revolutions per second. Therefore, the time required by light to travel a distance of 5.4 x 2 miles was $^1/_{25}$ x $^1/_{720}$ of a second. This gave a speed of 194,067 miles per second (the correct value is 186,300 miles per second).

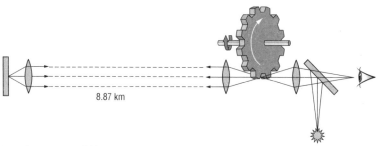

8.87 km

How Fizeau arranged his experiment.

1850

Second Law of Thermodynamics

Rudolf Clausius (1822–88)

Heat does not flow spontaneously from a colder to a hotter body.

There are many equivalent statements of the second law, each made by a different scientist at a different time.

The law says that many processes in nature are irreversible, never going backwards: burnt fuel is lost forever, an omelet cannot be turned back into eggs, isolated machines cannot stay in perpetual motion, and so on. It also defines the direction of time (time cannot go backwards).

In 1865, Clausius used the term **entropy** as a measure of the disorder or randomness of a system. The more random and disordered a system is, the greater the entropy. For example, ice has low entropy. Its entropy increases when it melts into water, and increases much more when water is heated and turned into steam. The entropy of an irreversible system must increase; therefore, the entropy of the universe is increasing.

The **third law of thermodynamics** dictates that it is impossible to cool an object to a temperature of absolute zero. This temperature is –273.15°C (–459.67°F) (see Absolute Zero, p. 92).

The American science-fiction writer John W. Campbell (1910–71) has provided the following interpretation of the laws of thermodynamics:

First law of thermodynamics:
You can't win.
Second law of thermodynamics:
You can't break even.
Third law of thermodynamics:
You can't quit the game.

1851

Foucault's Pendulum

FRANCE   Léon Foucault (1819–68)

A Foucault pendulum is a simple pendulum—a long wire with a heavy weight (bob) at the end—except that at the top it is attached to a joint which allows it to swing in any direction.

Foucault's pendulum proved that the Earth is rotating.

Once a Foucault pendulum is set in motion, it seems not to swing back and forth in the same direction but to rotate. In fact, it is the rotation of the Earth beneath the pendulum which gives rise to its apparent rotation. The angle of rotation per hour, which is constant at any particular location, can be calculated from the formula, $15 \sin q$, where q is the geographical latitude of the observer. At the North or South Pole, the pendulum would rotate through 360 degrees once a day. At the equator, it would not rotate at all.

In 1851, Foucault, an eminent physicist, gave a public demonstration of his pendulum at Le Panthéon, a church in Paris, with a 220-foot pendulum with a 62 pound cannon ball hung from the dome. A needle was attached to the ball and sand was sprinkled on the floor to trace out the ball's path. When he set the pendulum in motion, the needle slowly started to inscribe a figure like a twin-blade propeller in the sand, proving that the Earth was rotating on its axis.

1852

Frankland's Theory of Valency

Edward Frankland (1825–99)

The capacity of a given element to combine with other elements to form compounds is determined by the number of chemical bonds that element can form with other elements.

This "combining power" is now called valency (or valence). The concept of valency forms the foundation of modern structural chemistry.

In modern terms, valency is the number of electrons an atom of an element must lose or gain, either completely or by sharing, in order to form a compound. This leaves the atom with the stable electronic configuration of a noble gas (that is, a completely full outer shell). For example, in H_2O, hydrogen has a valency of $+1$ (H^+) and Oxygen -2 (O^{2-}). Two hydrogen atoms lose one electron each; one oxygen atom gains these two elec-

trons). Most elements have a fixed valency (for example, sodium $= +1$, chlorine $= -2$), but some have more than one valency (for example, iron $= +2$ or $+3$). The numerical values of valencies represent the charge on the ion.

Frankland, an organic chemist, not only gave chemistry students the concept of valency, he also introduced the word "bond" and the notation used commonly to represent the structure of compounds; for example, H–O–H. Frankland's concept of valency was not immediately accepted by his contemporaries. A few years later it was picked up and developed by Kekulé (p. 105).

Boolean Logic

George Boole (1815–64)

Logical operations can be expressed in mathematical symbols rather than words and can be solved in a manner similar to ordinary algebra.

Boole's reasoning founded a new branch of mathematics, now known as Boolean algebra, which has its own set of rules, laws, and theorems. Its most important application is in computer circuits and Internet search engines.

All computer circuits function in one of two states: on or off, which can be represented by 1 or 0. These two digits are known as binary digits or bits. Boolean algebra has three main logical operations: NOT, AND, OR. In NOT, for example, input is always the reverse of output. Thus NOT changes 1 to 0 and 0 to 1.

Boole was a self-taught mathematician and worked as a teacher in several village schools. When he published his first paper, "Mathematical Analysis of Logic," in 1847, he was offered the post of professor of mathematics at Queen's College, York, England. In 1854 he published his masterpiece, *An Investigation into the Laws of Thought*, which laid the foundation of Boolean algebra. Three years later he was elected a fellow of the Royal Society.

The importance of Boolean algebra was recognized when the first computers were built. Today's computers speak the language of 1s and 0s, the language invented by Boole, a boy from a working background who will always be remembered.

Darwin's Theory of Evolution

Charles Darwin (1809–82)

All present-day species have evolved from simpler forms of life through a process of natural selection.

Organisms have changed over time, and the ones living today are different from those that lived in the past. Furthermore, many organisms that once lived are now extinct.

Did you know about the reaction of the bishop's wife to the suggestion that man is derived from the apes? "Let's hope that is not true—or if it is, that it won't become generally known."

This is an example of the furor caused by the publication of Darwin's monumental book *On the Origin of Species by Means of Natural Selection* (often shortened to *The Origin of Species*) in 1859. All copies of the book were sold out on the first day of publication and it has been in print ever since. Many people strongly opposed the idea of evolution because it conflicted with their religious belief that every species was created by God in the form in which

it exists today and is not capable of undergoing any change. Darwin's theory continues to generate enormous social and scientific debate.

Darwin did not discuss the evolution of humans in his book. In a subsequent book, *The Descent of Man*, published in 1871, he presented his idea that humans evolved from apes.

In modern form the theory of evolution includes the following ideas:

- Members of a species vary in form and behavior and some of this variation has an inherited basis.

- Every species produces far more offspring than the environment can support.

- Some individuals are better adapted for survival in a given environment than others. This is called the "survival of the fittest." This means there are variations within each population gene pool and individuals with most favorable variations stand a better chance of survival.

- The favorable characteristics show up among more individuals of the next generation.
- There is thus a "natural selection" for those individuals whose variations make them better adapted for survival and reproduction.
- The natural selection of strains of organisms favors the evolution of new species, through better "adaptation" to their environment, as a consequence of genetic change or mutation.

Advances in modern biology, especially in knowledge of DNA, have enriched the theory of evolution. The modern view of evolution is still based on the Darwinian foundation: evolution through natural selection is opportunistic and it takes place steadily.

Kirchhoff Bunsen
Spectroscopy Theory

Robert Bunsen (1811–99)
Gustav Kirchhoff (1824–87)

Each chemical element, when heated to incandescence, produces its own characteristic lines in the spectrum of light.

For example, sodium produces two bright yellow lines.

Anyone who has ever worked in a school chemistry laboratory remembers the Bunsen burner. It was developed in 1855 by Bunsen, a great teacher and experimenter. In the **flame test**—a test to identify the presence of metals in a sample by the color of flames they produce—the Bunsen burner's non-luminous flame does not interfere with the colored flame given off by the sample.

When he was 24, Bunsen lost an eye in an experiment. He was a much-loved bachelor—"I never found time to marry"—of simple habits. The wife of a colleague at Heidelberg University, where Bunsen was a professor of chemistry, once declared that she would like to kiss Bunsen because he was such a charming man, but first she would need to wash him.

Bunsen was a friend of Kirchhoff, who was a professor of physics at Heidelberg. Bunsen and Kirchhoff together developed the first spectroscope, a device used to produce and observe a spectrum. They used their spectroscope to discover two new elements: cesium (1860) and rubidium (1861). However, in 1860 Kirchhoff made the momentous discovery that when heated to incandescence, each element produced its own characteristic lines in the spectrum. This means that each element emits light of a certain wavelength. In a leap of intuition he even went further: anything any atom emits it must also absorb. Sodium's spectrum has two yellow lines (wavelengths about 588 and 589 nanometers). The sun's spectrum contains a number of dark lines, some of which correspond to these wavelengths. This means that sodium is present in the sun. Scientists now had a tool they could use to determine the presence of elements in stars. According to Isaac Asimov (p. 168), Kirchhoff's banker was not impressed with his ability to find elements in the sun. "Of what use is gold in the sun if I cannot bring it down to Earth?" he asked. When Kirchhoff was awarded a gold medal for his work, he handed it to his banker and said: "Here's gold from the sun."

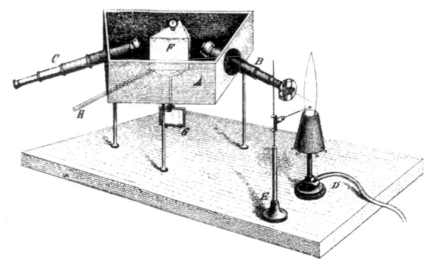

The apparatus used by Kirchhoff and Bunsen (diagram from *Annalen der Physik und der Chemie*, vol. 110, 1860).

See also Kirchhoff's Laws, p. 91.

1864

SCOTLAND

Maxwell's Equations

James Clerk Maxwell (1831–79)

Four equations that express mathematically the way electric and magnetic fields behave.

The equations also show that light is related to electricity and magnetism.

The equations are complex, but in simple words they describe: (1) a general relationship between electric field and electric charge; (2) a general relationship between magnetic field and magnetic poles; (3) how a changing magnetic field produces electric current; and (4) how an electric current or a changing electric field produces a magnetic field. The equations also predict the existence of electromagnetic waves, which travel at the speed of light and consist of electric and magnetic fields vibrating in harmony in directions at right angles to each other (see diagram).

Maxwell showed so little promise at school that he became known as "Dafty."

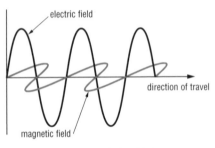

Diagram showing the relationship between a magnetic field and the changing electric field that produced it.

But when his father took him to a science lecture at the Royal Society of Edinburgh he soon became interested in science; he published his first scientific paper (on a technique for drawing a perfect oval) at the age of 14. Maxwell's scientific achievement is comparable with that of Newton or Einstein.

See also Kinetic Theory of Gases, p. 45.

1865 Mendel's Laws of Heredity

AUSTRIA   Gregor Mendel (1822–84)

Law of Segregation: **In sexually reproducing organisms, two units of heredity control each trait. Only one of such units can be represented in a single sexually reproductive cell.**
Law of Independent Assortment: **Each of a pair of contrasted traits may be combined with either of another pair.**

These laws laid the foundation for the science of genetics.

Mendel, a botanist and an Augustinian monk, used the edible garden pea plant for his research. His aim was to find the effect of cross-breeding on seven pairs of contrasting characteristics: seed shape, seed color, pod shape, pod color, flower color, flower position, and stem length. For seven years he kept an exact record of the inherited characteristics of some 28,000 pea plants; then he applied mathematics to these results. From this analysis he found that certain character-istics of plants are due to some factors passed intact from generation to genera-tion. These factors are now known as *genes*. He found that some factors were "dominant" and some were "recessive."

Mendel published his results in 1865, but they were soon forgotten. In 1900, sixteen years after Mendel's death, three European scientists independently started experiments on the heredity in plants and discovered that Mendel had already been there before. The monk and his peas have not been forgotten; they live on in all biology textbooks.

1865

Pasteur's Germ Theory of Disease

Louis Pasteur (1822–95)

Many human diseases have their origin in microorganisms.

This theory is one of the greatest scientific achievements of the nineteenth century.

Pasteur, a chemist, spent most of his life studying the origin of disease. In 1856, when he was a professor of chemistry and a dean at the University of Lille, he was approached by beetroot farmers who were having problems with the manufacture of alcohol from beetroot. After years of experimentation he proved that fermentation was caused by living microorganisms. In 1862 he showed that brief, moderate heating of wine and beer kills germs, thereby sterilizing them and ending the fermentation process. The process, now known as *pasteurization*, is still used in the food industry.

In 1865, these investigations led him to believe that microorganisms could also cause disease in humans. He was determined to find a way to kill these microorganisms. He isolated the germs that caused chicken cholera and discovered that in a culture these microbes produced a chemical which inhibited their growth. In 1881 he successfully applied his discovery of vaccination by attenuated culture of microorganism to anthrax, a disease of sheep and cattle, and then in 1885 to the treatment of rabies in humans. For the first time in history humans were winning against deadly germs. But it seems that the man who so boldly fought against germs was afraid of them: Pasteur avoided shaking hands for fear of infection.

1865

Kekulé's Theory of Organic Compounds

Friedrich August Kekulé (1829–96)

Carbon is tetravalent and is capable of forming ring-type organic molecules.

The concepts laid the foundation of structural chemistry.

In the 1860s chemists knew about the molecular formula of benzene, C_6H_6, but they did not know how the six carbon atoms were arranged in space. Even the German chemist Kekulé—who was the first chemist to suggest that carbon is tetravalent, that is, one carbon atom can combine with four atoms—was unable to work out the structure of benzene.

One day, when he was working as a professor of chemistry in Belgium, he was sitting dozing in front of a fire. He wrote later: "I turned my chair to the fire and dozed. Again the atoms were gamboling before my eyes. I could

distinguish long rows of them twisting and twining in snakelike motion. Then one of the snakes grabbed hold of its own tail and began rotating before my eyes." On awakening, Kekulé saw the possibility that the benzene molecule could be ring-shaped, as shown below.

A molecule of benzene.

1869 Mendeleev's Periodic Law

RUSSIA

Dmitri Mendeleev (1834–1907)

The properties of elements are periodic functions of their atomic weights.

Simply arrange the atoms in order of their atomic weight (now known as relative atomic mass) and elements are also arranged in order of their properties. This arrangement of the elements is called the periodic table.

In the modern periodic table elements are no longer arranged by atomic weight, but by a much more fundamental quantity: "atomic number." The atomic number of an element is the number of protons in the nucleus of one of its atoms; the number of neutrons, which contributes to atomic weight, is ignored. The modern periodic law is: The properties of elements are periodic functions of their atomic numbers.

"Yes, Mendeleev has two wives, but I have only one Mendeleev," replied Czar Alexander II of Russia when someone complained to him that Mendeleev had taken a second wife while not yet legally divorced from his first. In 1869, that only

one Mendeleev, at age 35, published a unique table of the 61 elements then known and predicted that some undiscovered elements would fill the gaps in the table. "It is possible to foretell the properties of still unknown elements," he wrote confidently.

His predictions were based on the properties of the elements surrounding the gaps in the table. He even gave them names—eka-aluminium, eka-boron, and eka-silicon (eka is a Sanskrit prefix for one). By 1886, his remarkable prediction was fulfilled, with the discoveries of gallium, scandium, and germanium; and they had properties Mendeleev had foretold. Mendeleev was suddenly the most respected chemist in the world. No wonder the czar held him in such high esteem.

Mendeleev's periodic law—the single most useful concept in chemistry—started the search for new elements. By 1925, chemists had successfully identified all the elements they believed to exist in nature. The first artificial element, neptunium, was synthesized in 1940. Many more elements have been made since then (p. 167).

The element Mendelevium, first created in 1955, perpetuates the memory of the chemist who set other chemists on the trail of hidden elements. And no chemistry classroom is complete without a periodic table.

Period	Group I	Group II	Group III	Group IV	Group V	Group VI	Group VII	Group VIII
1	H=1							
2	Li=7	Be=9.4	B=11	C=12	N=14	O=16	F=19	
3	Na=23	Mg=24	Al=27.3	Si=28	P=31	S=32	Cl=35.5	
4	K=39	Ca=40	?=44	Ti=48	V=51	Cr=52	Mn=59	Fe=56 Ca=59
								Ni=59 Cu=63

English translation of a section of Mendeleev's 1869 periodic table; note the missing element at atomic weight 44. It was discovered in 1879 in Sweden and is named scandium (Sc).

1879 Stefan–Boltzmann Law

AUSTRIA

Josef Stefan (1835–93)
Ludwig Boltzmann (1844–1906)

**The total energy radiated from a blackbody is proportional
to the fourth power of the temperature of the body. (A blackbody is
a hypothetical body that absorbs all the radiation falling on it.)**

The law was discovered experimentally by Stefan, but Boltzmann discovered it theoretically soon after.

The law has many practical applications, but an unusual application appeared in an unsigned article "Heaven is hotter than Hell" in *Applied Optics*, vol. 11 (1972). The article starts with a quote from the *Bible*: Isaiah 30:26: "The light of the moon shall be as the light of the sun and the light of the sun shall be sevenfold, as the light of seven days."

Thus "Heaven" receives from the moon as much as radiation as we do from the sun and in addition seven times seven (49) times as much as the Earth does from the sun, or 50 times in all. The radiation reaching Heaven will heat it to the point where the heat lost by radiation is just equal to the heat received by radiation. Or, Heaven loses 50 times as much heat as does the Earth. From the Stefan–Boltzmann law the Earth's temperature is 525°C (977°F). According to Revelation 21:8—"But the fearful and unbelieving … shall have their part in the lake which burneth with fire and brimstone"—the temperature of Hell must be less than 445°C (833°F), the temperature at which brimstone or sulfur changes into a gas. Therefore, Heaven is hotter than Hell.

1883   Reynolds Number

ENGLAND

Osborne Reynolds (1842–1912)

The ratio of pressure forces to viscosity forces in a fluid flow.

The Reynolds number is a dimensionless quantity (that is, it has no units). The number has great importance in fluid dynamics.

The number depends upon the speed, density, viscosity, and linear dimensions (such as diameter of a pipe or height of a building) of the flow. Fluid flow is described as "turbulent" when the number is greater than 2000. It is considered "laminar" (steady) when the value is less than 2000.

Reynolds, a renowned theoretical engineer, presented the concept of a number to determine the type of a fluid flow in a paper with a very long title—"An experimental investigation of the circumstances which determine whether motion of water shall be direct or sinuous and of the law of resistance in parallel channels"—in the *Philosophical Transactions of the Royal Society*. In this paper he observed that 'the tendency of water to eddy becomes much greater as the temperature rises." He associated temperature rise with a decrease in viscosity (the resistance of a fluid to flow).

In 1868 Reynolds became the first professor of engineering at the University of Manchester, UK, a post he was to hold until his retirement in 1905. In 1877 he was elected a Fellow of the Royal Society. Eleven years later he was awarded the society's Royal Medal.

1884

SWEDEN

Arrhenius' Theory of Ionic Dissociation

Svante Arrhenius (1859–1927)

When an ionic compound such as sodium chloride is dissolved in water, the electrostatic attraction between its positive and negative ions becomes very weak and the ions separate. For example, sodium chloride, NaCl, separates into Na^+ and Cl^- ions.

This process is known as ionic dissociation. The process has practical application in electroplating and other industrial processes.

Once the ions are dissociated, an electric current can pass through the solution and the solution is known as an electrolyte. Positive ions such as Na^+ are attracted to the negative pole (cathode) and negative ions such as Cl^- to the positive pole (anode). The ions of ionic compounds also become separated when these compounds are melted. That's why ionic compounds conduct electricity in the molten state as well.

Arrhenius also said that an acid is a substance that in water solution gives hydrogen ions, H^+; and a base is a substance that gives hydroxide ions, OH^-. He described acid–base neutralization as a process in which H^+ and OH^- join to form a water molecule, H_2O. **Arrhenius' concept of acids and bases** is very narrow in scope as it only applies to water solutions of acids and bases. It was replaced in 1923 by the Brønsted–Lowry concept of acids and bases (p. 143).

See also Greenhouse Effect, p. 123.

1887

US

Michelson–Morley Experiment

A. A. (Albert Abraham) Michelson (1852–1931)
Edward Morley (1838–1923)

The aim of the experiment was to measure the effect of the Earth's motion on the speed of light.

This celebrated experiment found no evidence of there being an effect.

In the nineteenth century scientists believed that light waves traveled through "ether" which filled space. As the Earth moves in its orbit at a speed of about 18.6 miles per second, scientists expected some effect of the Earth's motion on the speed of light. In other words, light waves traveling in the direction of the Earth's motion would move faster, being helped by the motion of the hypothetical "ether," while the light waves traveling in the direction opposite to the Earth's motion would be slowed down.

Michelson and Morley's experiment was designed to detect any change in the speed of light. In their experiment, performed over four days in a cellar in Cleveland, Ohio, they split a beam of light into two and sent them backwards and forwards between mirrors in two directions at right angles. The apparatus was mounted on a massive stone floating in mercury, so that it could be rotated in any direction. They found that the speed of light was always the same, independent of the speed of its source. They could not explain the discrepancy. The explanation came in 1905 from Einstein in his special theory of relativity (p. 130).

1888 Hertz's Radio Waves

GERMANY   Heinrich Hertz (1857–94)

Radio waves can be produced by electric sparks. They have the same speed as light and behave the same as light.

Hertz's discovery provided the basis of radio broadcasting. Radio waves are electromagnetic waves. Other main kinds of electromagnetic waves are: gamma rays, x-rays, ultra-violet radiation, visible light, infrared radiation, and microwaves.

In 1864 Maxwell's equations (p. 102) showed the existence of electromagnetic waves, but no one could find a way to prove their existence. In 1888 Hertz, a newly married young professor at the Technical University of Karlsruhe, set out to find the elusive waves. He modified an induction coil to generate sparks across a gap between two brass balls. In those days it was a common setup for demonstrating electric discharge. A few yards away from the induction coil he placed a loop of wire connected to two brass balls separated by a tiny gap. When he passed the discharge through the induction coil he was startled to see a tiny spark in the loop of wire a short

distance away. His wife was with him to witness the first radio transmission in history: one wave emitted at one point had been received at another. No music, no talk-back shows, just a tiny blue spark and an entry in Hertz's diary for November 1, 1886: "Vertical electric vibrations, in wires stretched in straight lines, discovered; wavelength, 3 meters."

The induced spark proved that electromagnetic waves do exist. A year later Hertz was able to measure the speed of these waves and to show that the speed was the same as that of light. His further experiments showed that electromagnetic waves could be refracted, reflected, and polarized in the same way as light.

Hertz never realized the importance of his discovery. When he demonstrated his experiment to his students, someone asked what the discovery could be used for. "Nothing, I guess," replied Hertz. It was left to the Italian physicist Guglielmo Marconi (1874–1937) to develop technology for the practical use of Hertzian waves.

We are now familiar with all the types of electromagnetic waves that make up the complete electromagnetic spectrum. They all travel with the speed of light and differ from each other in their frequency—and we measure this frequency in hertz (Hz), a unit named in Hertz's honor.

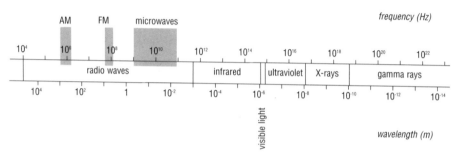

A part of the electromagnetic spectrum (from radio waves to gamma rays).

1888

FRANCE

Le Châtelier's Principle

Henri Louis Le Châtelier (1850–1936)

When a system in equilibrium is subjected to a change in conditions, it adjusts itself so as to try to oppose that change.

The principle is a consequence of the law of conservation of energy (p. 88).

"Reflection will show that this principle necessarily follows if we are to have conservation of energy; otherwise perpetual motion machines will be a dime a dozen, and we would be able to create an unlimited amount of energy from any small starting push," George Gamow (p. 171) said in 1963. Perpetual motion machines still fascinate people and if you key in the words in your Web search engine, thousands of sites will appear on your screen to entertain you perpetually.

Le Châtelier's principle is invaluable in understanding how to control the industrial production of chemicals such as ammonia. In the production of ammonia, nitrogen and hydrogen react to form ammonia, that is, nitrogen + hydrogen $\rightleftharpoons$ ammonia (the double arrow shows that it's a two-way reaction). When the pressure of this system is increased, more ammonia is produced, but when the pressure is lowered, ammonia is decomposed into nitrogen and oxygen. Thus, by controlling pressure and temperature chemists can produce ammonia with a minimum of waste.

Le Châtelier was a chemist and taught at the École des Mines in Paris. He is also remembered for inventing thermocouples for measuring high temperatures (1877), and oxyacetylene welding (1895).

1888

Tesla's Concept of Alternating Current

Nikola Tesla (1856–1943)

The transmission of high-voltage alternating current (AC) over long distances is more efficient than the transmission of direct current (DC).

DC transmission is no longer used anywhere in the world.

In the 1880s the American inventor Thomas Edison (1847–1931) developed DC generation and set up his Edison Light Company to build power plants. DC loses much of its energy when transmitted through wires over long distances and, therefore, DC power plants had to be close to cities. In 1888 Tesla invented the AC motor and suggested that transmission of AC is more efficient. Edison fiercely opposed Tesla's ideas, but Tesla persuaded the entrepreneur George Westinghouse (1846–1914) to build the first commercial AC power plant at Niagara Falls. On November 16, 1896 this hydroelectric power plant became the first power plant to transmit electric power between two cities (from Niagara Falls to Buffalo, New York).

The Croatian-born Tesla was a genius so ahead of his time that his contemporaries failed to understand his groundbreaking inventions. Besides inventing and developing AC power, he invented induction motors, dynamos, transformers, condensers, bladeless turbines, mechanical rev counters, automobile speedometers, gas-discharge lamps that were the forerunners of fluorescent lights, radio broadcasting, and hundreds of other things (the number of patents in his name exceeds 700). A unit for measuring magnetism, tesla (T), is named in his honor.

1889 Friese-Greene's Magic Box

ENGLAND   William Friese-Greene (1855–1921)

Friese-Green's magic box was a camera that was capable of taking a series of photographs on a roll of perforated film moving intermittently behind a shutter.

This is the basic principle of the motion picture camera.

Photographer and inventor Friese-Greene experimented with moving pictures in the early 1880s, long before Thomas Edison's invention of the kineto-scope later that decade. This peepshow machine had an eyepiece for the viewer and showed cardboard pictures in quick succession, thus creating the effect of continuous motion.

In the beginning of his professional career, Friese-Greene worked with J. A. R. Rudge, inventor of the bio-phantascope, an improved magic lantern which used a series of slides in quick succession. Later on, he opened his photography shop in Piccadilly, London. There he spent his time experimenting. His main object was to find the right kind of film. First he tried plate glass, then paper strips soaked in castor oil to make them transparent, and finally celluloid paper coated with sensitive emulsion. In celluloid paper he found the kind of film he was searching for.

Friese-Greene was not a good mechanic. With the help of an engineer friend, he built his movie camera. The camera was capable of taking a series of photographs on a roll of perforated film moving intermittently behind a shutter.

One morning in 1889, he went out into London's Hyde Park to shoot about 20 feet of film. That night, in his laboratory, he developed and printed a strip and ran it through his projector. Immediately he saw on the screen a life-long dream becoming true—hansom cabs and jerky pedestrians, all moving almost as in real life! He was almost insane with excitement at the discovery.

"I've got it! I've got it!" Shouting with joy, Friese-Greene ran into the street and dragged a passing policeman into his house. He ran through the film again on his projector for his startled audience of one, making him the first movie watcher in the world. That night, a dream changed into reality for Friese-Greene. He was the first man to shoot and show a motion picture.

This joy in the poor inventor's life was not to last. He could not persuade any financier to back his invention. In the nineteenth century, no invention was attempted by so many inventors as that of moving pictures. For this reason, it is not easy to say which of them really created cinematography. However, Edison is hailed as the inventor of cinematography, because Friese-Greene failed to give a successful presentation of moving pictures.

Friese-Greene died penniless at the age of 66. The epitaph on his tomb in Highgate Cemetery, London, reads: "His genius bestowed upon humanity the boon of commercial photography." John Boulting's marvelous 1951 movie *The Magic Box* (starring Robert Donat as Friese-Greene and Laurence Olivier as the policeman Friese-Greene pulls off the street) is based on the life of this forgotten inventor of the movies.

1890

ENGLAND

Fleming's Rules

John Ambrose Fleming (1849–1945)

Fleming's left-hand and right-hand rules are used for the relationship between the directions of current flow, motion, and magnetic field in electric motors and dynamos, respectively.

The rules are useful mnemonics for students.

The rules, illustrated below, are named after the English electrical engineer,

inventor of the thermionic valve (1904), who devised them.

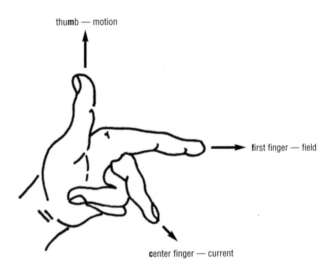

thumb — motion

first finger — field

center finger — current

Fleming's left-hand rule is used for electric motors. Hold left hand with the thumb, first, and second fingers at right angles. Directions are as indicated. Use the right-hand for dynamos. This is called Fleming's right-hand rule.

1894

IRELAND

1904

HOLLAND

Lorentz–Fitz Gerald Contraction

George Fitz Gerald (1851–1901)
Hendrik Lorentz (1853–1928)

A moving object appears to contract.

The contraction is negligible unless the object's speed is close to the speed of light.

In 1894 Fitz Gerald suggested that an object moving through space would shrink slightly in its direction of travel by an amount dependent on its speed. In 1904 Lorentz independently studied this problem from an atomic point of view and derived a set of equations to explain it. From these equations, for example, it can be calculated that a 39-inch ruler moving past us at a speed of 149,100 miles per second (that is, 80 percent of the speed of light) would seem only 23.6 inches long to us, provided we are not moving with it.

The idea of contractions of length with motion led an unknown wit to write the following famous limerick:

There was a young fellow named Fisk
Whose fencing was exceedingly brisk.
So fast was his action,
The Fitz Gerald contraction,
Reduced his rapier to a disk.

A year later Einstein derived Lorentz's equations independently, from his (Einstein's) special theory of relativity (p. 130).

1894

Ostwald's Principle of Catalysis

Friedrich Wilhelm Ostwald (1853–1932)

A catalyst can change the rate of a chemical reaction, but is not itself used up in the reaction.

The effect of a catalyst is known as catalysis. The action of a catalyst is specific (that is, a particular catalyst is necessary to catalyze a given reaction) and it can increase or decrease the rate of a reaction.

The word "catalysis" comes from the Greek word *katalusis* meaning "to loosen up" and was first used by Jöns Jacob Berzelius (p. 86) in 1836. Ostwald, a physical chemist, gave the first modern definition of catalysis and made a detailed study of catalytic reactions. He was awarded the 1909 Nobel Prize for chemistry for his work on catalysis.

There would be no life without catalysts. Living cells produce enzymes which act as catalysts for all the chemical reactions that take place in living organisms. Ostwald was the first to prove the catalytic action of enzymes. The human body contains thousands of different kinds of enzymes. For example, an enzyme called amylase in our saliva increases the rate at which starch is converted to sugars.

Air pollution is reduced due to the use of a catalyst in your car: a device known as a catalytic converter speeds up reactions in which noxious unburnt fuel, carbon monoxide, and nitrogen oxides are converted into environmentally friendly carbon dioxide, water, and nitrogen.

1895

Röntgen's X-Rays

Wilhelm Röntgen (1845–1923)

X-rays are high-energy radiation given off when fast-moving electrons lose energy very rapidly.

X-rays are highly penetrating and in large doses can cause serious damage to living tissue. The discovery initiated the modern age of physics and revolutionized medicine.

Röntgen, a professor of physics at the University of Würzburg, discovered x-rays accidentally when he was experimenting with a Crooke's tube (a high-voltage gas discharge tube). He noticed that a little barium platinocyanide screen lying on the bench glowed brightly when he passed the current through the tube. When he placed his hand between the screen and the tube he was surprised to see an image of the bone on the screen. He wrapped a photographic plate and a key in brown paper and placed the packet near the tube. On developing, a silhouette of the key appeared on the photographic plate.

Röntgen worked feverishly and secretly—he did not even tell his wife about his discovery—for seven weeks to find out the properties of the mysterious radiation coming out of the tube. He announced his discovery in a paper "On a New Kind of Rays, a Preliminary Communication" in which he methodically set down in 17 numbered sections the properties of the new rays: they can pass through wood, paper, and aluminum; they can ionize gases; they are not affected by electric and magnetic fields; and they do not exhibit any properties of light. He named them x-rays.

"Then hell broke lose," Röntgen wrote later. The news caused an immense sensation around the world. While scientists applauded the new discovery and worked on Röntgen's experiments, charlatans worked on the gullible public selling x-ray proof underwear and other devices. The newspapers of the time were more interested in reporting fanciful speculation than scientific facts. A sample: "It is suggested that, if all that has reached us by cable is true, there will no longer be any privacy in a man's house, as anyone with a vacuum tube outfit can obtain a full view of any interior through a brick wall." And another:

I hear they'll gaze
Through cloak and gown—and even
   stays,
These naughty, naughty Roentgen rays.

Even the esteemed *Scientific American* (February 22, 1896), instead of reporting on the scientific side of the story, resorted to reprinting a poem from the London *Punch*. An excerpt:

O Roentgen, then the news is true
   And not a trick of idle rumor
That bids us each beware of you
   And of your grim and graveyard
      humor.

Fortunately, this silliness did not last for long. Within a few months x-rays were being used for medical diagnostic work. Röntgen did not patent his apparatus. He was awarded the first Nobel Prize for physics in 1901.

Greenhouse Effect

Svante Arrhenius (1859–1927)

Energy radiated from the surface of the Earth is absorbed by carbon dioxide, acting as a thermal blanket around the globe, thus creating the greenhouse effect.

A century after Arrhenius proposed his theory, we now know that the greenhouse effect is caused by many heat-trapping gases—carbon dioxide, nitrous oxide, methane, ozone, and fluorocarbons—present in the atmosphere.

Burning of fossil fuels and destruction of forests continue to increase the concentration of carbon dioxide in the atmosphere. Carbon dioxide is the most abundant of the greenhouse gases. The concentrations of nitrous oxide (main source: vehicle exhausts), methane (main source: cattle flatulence; for example, a cow belches about 300 liters of methane every day) and fluorocarbons (main source: industrial processes) are also increasing. These gases absorb the Earth's radiation and cause **global warming**.

The global average surface temperature has increased by 0.34°F since the greenhouse effect was predicted by Arrhenius. If global warming remains uncontrolled, scientists predict that global surface temperature could rise by up to 4.5°F in the next 50 years, with significant regional variations. Rising global temperatures are expected to raise sea level, and change precipitation and other local climate conditions. Our planet, on average, will be wetter and more humid.

See also Arrhenius' Theory of Ionic Dissociation, p. 110.

Thomson's Model of the Atom

Joseph John Thomson (1856–1940)

The atom is a sphere of positively charged protons in which negatively charged electrons are embedded (much like raisins in a dessert) in just sufficient quantity to neutralize the positive charge.

Thomson's model was the first model for the internal structure of the atom.

In 1886 the German physicist Eugen Goldstein (1850–1931) discovered that a cathode ray tube emits streams of positively charged particles as well as the usual cathode rays. These particles were named protons. In the following year Thomson showed that cathode rays could be deflected by a magnetic or an electric field. He concluded that cathode rays were streams of negatively charged particles and that these particles (later named electrons) came from the atoms of the metal of the negatively charged electrode, or cathode.

It was determined in 1909 that the mass of an electron equals $1/1837$ of that of a proton (p. 133). Electrons are the fundamental units of electricity: electricity is a flow of electrons in a conducting medium. The discovery of the proton and the electron led Thomson to propose his famous model of the atom. Thomson's model was soon replaced by a superior model advanced by his student Rutherford (see p. 136).

1897

Landsteiner's Concept
of Blood Groups

Karl Landsteiner (1868–1943)

**Each person has blood that falls into one of
three "blood groups": A, B, and O.**

The fourth blood group, named AB, was discovered a year later. The discovery of blood groups made blood transfusion safe.

Landsteiner, an assistant at the University of Vienna's Institute of Hygiene, became interested in "differences" in blood when he saw many people die from blood transfusion during surgery. For this remarkable insight he was awarded the 1930 Nobel Prize for physiology and medicine.

Now, human blood is classified on the basis of the presence or absence of certain antigens on the surface of the red cells and the natural antibodies (p. 181) in the plasma. Blood transfusion is safe only if blood groups are compatible.

Blood group	Antigen on red blood cells	Antibody in plasma	Can give blood to	Can accept blood from
A	A	anti-B	A and AB	A and O
B	B	anti-A	B and AB	B and O
AB	A and B	neither	AB	all groups
O	neither	anti-A and anti-B	all groups	O

The Curies' Experiments on Pitchblende

Marie Curie (1867–1934)
Pierre Curie (1859–1906)

Pitchblende, the ore from which uranium is extracted, is much more radioactive than pure uranium. The ore must therefore contain unknown radioactive elements.

The Curies isolated two new radioactive elements, polonium and radium, from pitchblende.

In 1896 the French scientist Henri Becquerel (1852–1908) accidentally left a specimen of pitchblende—the peaty brown natural ore from which uranium is obtained—on a photographic plate which was covered with a protective black paper. When he developed the plate by mistake, to his surprise, he discovered an image on it. The image looked like the container in which pitchblende had been stored. Becquerel carefully repeated this accidental experiment and showed that pitchblende emitted invisible radiation. But he was unable to explain the nature of this radiation.

Becquerel's discovery attracted the attention of Polish-born Marie and her husband, French-born Pierre. Marie systematically investigated the phenomenon and showed that the amount of radiation from pitchblende was proportional to the weight of uranium present.

She was convinced that pitchblende also contained other radioactive elements, as it gave off more radiation than could be accounted for by uranium. Pierre gave up his own research work to help Marie in her search for the new element. Their experiments were conducted in a miserable old shed at the School of Physics in Paris. The shed was cold and wet in winter, unbearably hot in summer. It was always filled with smoke and fumes from a bubbling cauldron. After four years of arduous work—which involved separating the constituents of six tons of pitchblende by chemical and physical experiments— Marie eventually succeeded in extracting a few milligrams of two new elements.

She named the first element polonium (after her native Poland), the second radium (from the Latin *radius*, ray), and the emission of radiations by certain substances "radioactivity."

In 1903 Marie shared the 1903 Nobel Prize for physics for studies on radioactivity with Pierre and Becquerel. She was awarded the 1911 Nobel Prize for chemistry for her discovery of polonium and radium. Marie was not only a great scientist but a great woman too. Her life is an example of true courage and perseverance. Her discoveries were for the benefit of humanity. She refused to take out extremely profitable patents on radium, saying that radium is an instrument of mercy and belongs to the world.

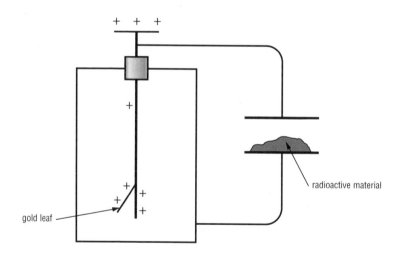

The Curies used a simple gold-leaf electroscope for testing radioactivity. When the electroscope is given a charge, the gold leaf is repelled. When a radioactive substance is placed on the plate, the charge leaks away and the gold leaf falls.

1900

Quantum Theory

Max Planck (1858–1947)

**Energy is not a continuous quantity but it is quantized:
that is, it flows in discrete bundles or quanta (singular quantum).
When particles emit energy they do so only in quanta.**

The theory marked the birth of quantum mechanics (the study of the motion and interaction of elementary particles).

According to the quantum theory, the energy (E) of one quantum (now called a *photon*) is given by $E = hf$, where f is the frequency of radiation and h is a constant, now called **Planck's constant**, and its value is 6.63 x 10^{-34} joules per second.

Quantum theory was immediately accepted by scientists. In 1905 Einstein applied it in his theory of photoelectric effect. Bohr used it in 1913 to develop the quantum model of the atom (p. 139). Planck was awarded the 1918 Nobel Prize for physics.

Though he was always antiNazi, Planck felt he was too old to oppose Hitler's rise. In Hitler's Germany, he continued to work as the president of Kaiser-Wilhelm Institute (now Max Planck Institute). A few months before he died he wrote a detailed account of how he sought an audience with Hitler "to put in a word in favor" of his Jewish colleagues. His friend Einstein never forgave him for not showing firmer opposition to Hitler, and never communicated with him again postwar.

1903

RUSSIA

Pavlov's Theory of Conditioned Reflexes

Ivan Pavlov (1849–1936)

An innate or built-in reflex is something we do automatically without thinking (such as moving our hand away from a flame). A conditioned reflex is a learnt response to an environmental stimulus (for example, a bell could, through experience, trigger a dog to drool). The process of learning to connect a stimulus to a reflex is called conditioning.

Pavlov's work paved the way for an objective study of human behavior.

In the 1890s, Pavlov, a physiologist, did exhaustive studies of the digestive system of dogs. He was awarded the 1904 Nobel Prize for physiology or medicine for his pioneering work. However, Pavlov started his best known work on conditioned reflexes in 1903 and continued working on it until his death.

During his experiments on dogs, Pavlov found that without salivation, the brain did not get the message to start digesting. He wondered whether the learning process could affect salivation. To test this, he started a series of experiments on dogs which is now known to every psychology student. In one experiment, he played a metronome just before giving a dog some food. After a while, the dog started dribbling saliva in response to the sound of the metronome even before the food arrived. From his experiments Pavlov concluded that innate and conditioned reflexes permitted an animal to modify its behavior on the basis of experience.

1905 Special Theory of Relativity

SWITZERLAND

Albert Einstein (1879–1955)

> **(1)** *The relativity principle*: **All laws of science are the same in all frames of reference.**
> **(2)** *Constancy of the speed of light*: **The speed of light in a vacuum is constant and is independent of the speed of the observer.**

These are the two fundamental assumptions of the theory.

When he formulated his special theory of relativity Einstein was 26 years old and working as a clerk at the Swiss Patent Office in Berne. The significance of Einstein's theory was not recognized by scientists for many years as there was no experimental evidence to support it.

The theory says that time is not an absolute quantity. Our measurements of time are affected by our motion. The rate at which clocks run depends upon their relative motion. A person running away from a clock would observe it to move more slowly than his or her own clock. The theory also says that the mass of a moving object increases as its speed increases. At the speed of light, which is about 186,300 miles per second, the mass becomes infinite and therefore nothing can move faster than light.

Theoretically, a spaceship traveling nearly at the speed of light would take nine years by the Earth's calendar to make a return trip to Centauri, the next nearest star after the sun; but because of relativistic time changes, on return to the Earth the crew would find that many decades have gone by. However, the crew would notice no change on the spaceship. From their point of view, the spaceship is stationary and the Earth is moving at almost the speed of light and time on the Earth slows down.

Relative time poses an interesting paradox: if one twin goes on a high-speed space journey, she would return younger than her sister who stayed home.

The special theory of relativity defies common sense, but to Einstein common sense was just a "deposit of prejudice laid down in the mind prior to the age of eighteen." He once remarked that it was the common sense which once objected to the idea that the Earth is round.

And if you could travel much faster than light, as Miss Bright did according to the famous limerick, you would return home the day before you had left.

There was a young girl named Miss Bright
Who could travel much faster than light.
She departed one day,
The Einsteinian way,
And came back the previous night.

As long as Einstein's theory is supreme, time travel into the past would remain in the realm of science fiction. Meanwhile, Virgil, the greatest of Latin poets, remains true: "Time is flying—flying never to return."

See also Theory of General Relativity, p. 140.

E = mc²

Albert Einstein (1879–1955)

The energy of a body (*E*) equals its mass (*m*) times the speed of light (*c*) squared.

The world's most famous equation shows that mass and energy are mutually convertible under certain conditions.

For centuries scientists believed that energy and mass were two entirely separate things. Einstein showed that mass and energy were one. The mass–energy equation is a consequence of Einstein's theory of special relativity (p. 130).

To appreciate the significance of $E = mc^2$, consider the following: E is energy in joules, m is mass in kilograms and c is the speed of light in metres per second. Thus, the energy released by 1 kilogram of matter

= 1 x 300,000,000 x 300,000,000 joules

= 90,000 million million joules

= energy released by 20,000 kilotons of TNT.

The Hiroshima atomic bomb was only a 15-kiloton bomb. For years Einstein believed that energy could not be released on such a tremendous scale. But the Hiroshima atomic bomb proved him wrong. We can say that $E = mc^2$ ushered in the atomic age.

Millikan's Oil-Drop Experiment

Robert Millikan (1868–1953)

Millikan measured the charge on the electron.

The experiment showed that the electron was the fundamental unit of electricity: that is, electricity is the flow of electrons.

The ingenious experimental arrangement used by Millikan consisted of a small box attached to a microscope. An atomizer was used to introduce mineral oil drops between two charged circular plates, which were ¾ inch apart. By adjusting the voltage, the charge on the plates was changed until the drops were suspended in mid-air. For this condition, the charge on the oil drop (upward electric force) overcame its weight (downward pull of gravity).

From his experiment Millikan calculated the basic charge on an electron to be 1.6×10^{-19} coulomb (by convention this charge is called unit negative, -1, charge). This charge cannot be subdivided. Millikan also determined that the electron has only about $1/1837$ the mass of a proton, or 9.1×10^{-31} kilogram.

In 1923 Millikan became the second American (Michelson, p. 113, had been the first, in 1907) to win the Nobel Prize for physics. He once said: "Cultivate the habit of attention and try to gain opportunities to hear wise men and women talk. Indifference and inattention are the two most dangerous monsters that you ever meet. Interest and attention will ensure to you an education."

pH Scale

Søren Peter Sørensen (1868–1939)

A scale of acidity and alkalinity. It runs from 0 (most acid) to 14 (most alkaline). A neutral solution has a pH of 7. A solution is acidic when the pH is less than 7, and basic (alkaline) when the pH is greater than 7.

The scale is logarithmic; for example, a glass of beer with a pH of 4 is ten times more acidic than a cup of black coffee with a pH of 5.

The pH (short for "power of hydrogen") measures the concentration of hydrogen ions, H–, in water. Therefore, the pH scale can only be used for solutions of acids and bases in water. Some common pH values are shown below:

Battery acid	0.1 to 0.3	Most drinking water	6.3 to 6.6
Stomach acid	1.0 to 3.0	Pure water	7.0
Vinegar	2.4 to 3.4	Seawater	7.8 to 8.3
Soft drinks	2.5 to 3.5	Ammonia	10.6 to 11.6
Soil (best for most plants)	6.0 to 7.0	Drain cleaner	14

1911 Superconductivity

HOLLAND Heike Kamerlingh Onnes (1853–1926)

At very low temperatures, some materials conduct electricity without any resistance: that is, virtually without any loss of energy.

These materials are called super-conductors; they have many technological applications.

In 1908 Kamerlingh Onnes, a physicist, managed to cool helium to temperatures close to absolute zero ($-459.67°F$, p. 92). He used this "refrigerator" to study properties of metals at low temperatures. In 1911 he found that metals such as mercury, lead, and tin become super-conductors at very low temperatures. He was awarded the 1913 Nobel Prize for physics for his discovery.

Now scientists know that about 24 elements and hundreds of compounds become superconductors near absolute zero, a temperature that can only be achieved with helium. Because helium is rare and expensive,

superconducting technology advanced little until 1986, when scientists developed a metallic ceramic compound that becomes superconductive at around the temperature of liquid nitrogen, $-321°F$.

The most important medical application of superconductors is in magnetic resonance imaging (MRI), which is used by doctors to probe the human body. MRIs work on extremely powerful electromagnets. These magnets, if made from ordinary metals, would be as big as a truck and would generate so much heat that literally rivers of water would be needed to cool them. Super-conducting electromagnets do the same job without generating any heat. And they are so small that they can fit on a coffee table.

Rutherford's Model
of the Atom

Ernest Rutherford (1871–1937)

The atom contains a core or nucleus of very high density and very concentrated positive charge. Most of the atom is empty space, with the electrons, like the planets around the sun, moving about the tiny central nucleus.

Rutherford's model is still regarded as essentially correct, but it was modified by later scientists.

New Zealand-born Rutherford was professor of physics at McGill University in Canada when in 1902 he announced his **theory of radioactivity**: "Radioactivity was due to nuclear disintegration, that is, breaking up of the nucleus of the atom into smaller parts. A radioactive atom emits three types of radiation—alpha particles, beta particles, and gamma rays—and decays into a new element." Although the theory initially met with widespread incredulity (one of his colleagues asked sarcastically why it was that radioactive atoms were seized with "an incurable suicidal mania"), Rutherford was awarded the Nobel Prize for chemistry in 1908 for his discovery.

He delighted in telling friends that the fastest transformation he knew of was his transformation from a physicist to a chemist.

In 1909, now at Manchester University in England, Rutherford asked his colleague, Hans Geiger, who later invented the celebrated Geiger counter, and Ernest Marsden, a brilliant graduate student, to study the scattering of alpha particles, which are positively charged helium nuclei. They bombarded a thin gold foil with high-velocity alpha particles from a radioactive element. Almost all the particles passed straight through the gold foil (which was about 1000 atoms thick) but about one in every 20,000 bounced back. According to calculations based on Thomson's model of the atom (p. 124), positively charged alpha particles were expected to pass directly through the gold foil just as a bullet will go through a sheet of paper.

The result of the experiment astonished Rutherford. "It was almost as incredible as if you fired a 15-inch shell at a piece of tissue paper and it came back," he remarked. To account for the experimental observations Rutherford suggested that most of the gold atom was empty space and almost all of the mass and positive charge were concentrated in a minute nucleus. The majority of the alpha particles missed the tiny gold nuclei and passed through the foil without hindrance.

Rutherford suggested a new model of the atom: "The atom contains a core or 'nucleus' of very high density and very concentrated positive charge. Most of the atom is empty space with the electrons, like the planets round the sun, moving about the tiny central nucleus." During the next 12 months, Geiger and Marsden conducted more experiments which proved that the new model was indeed correct. After announcing his theory of the atom in 1911, Rutherford, with a broad grin, said to his close colleagues of his critics: "Some of them would give a thousand pounds to disprove it."

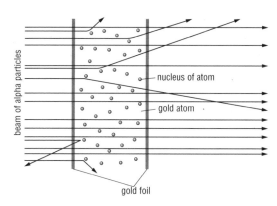

How gold foil revealed the existence of the atomic nucleus. The diagram shows how some alpha particles pass straight through the gold foil, while others are deflected or bounce back after coming into contact with gold nuclei.

See also Bohr Atom, p. 139.

1912

Bragg's Law

William Henry Bragg (1862–1942)
William Lawrence Bragg (1890–1971)

X-rays scattered from a crystal will show constructive interference provided their wavelength l fits the equation, $2d\sin q = nl$, where d is the spacing between atoms of the crystal, q the angle through which the rays have scattered, and n is any whole number.

Complex it may be, but the law is the cornerstone of the clever science of crystallography, which has opened up the amazing microscopic world of atoms and molecules.

William Henry and William Lawrence were father and son. William Henry was born in England and came to Australia in 1885 as Professor of Mathematics and Physics at the University of Adelaide. His elder son, William Lawrence, was born in Adelaide and graduated from the University of Adelaide. In 1909 the Braggs returned to England and there father and son worked together on crystal structures using x-rays, thus founding the science of x-ray crystallography.

For their pioneering work on x-rays and atomic structure of crystals, and their discovery that atoms in crystals are arranged in regular formations, like oranges stacked on a fruit stall, they were jointly awarded the 1915 Nobel Prize for physics. The Braggs are the only father-and-son pair ever to have been honored in this fashion. William Lawrence is still the youngest person (at 25 years old) to have won the Nobel Prize.

Bohr Atom

Niels Bohr (1885–1962)

Electrons in atoms are restricted to certain orbits, called "allowed orbits," but they can move from one allowed orbit to another.

Bohr's model was the first quantum model for the internal structure of the atom.

Bohr, who worked with Rutherford in Manchester, improved upon Rutherford's model (p. 136), which said that electrons were free to orbit the nucleus willy-nilly. In addition to showing that electrons are restricted to orbits, Bohr's model also suggested that:

- the orbit closest to the nucleus is lowest in energy, with successively higher energies for more distant orbits
- when an electron jumps to a lower orbit, it emits a photon
- when an electron absorbs energy, it jumps to a higher orbit.

Bohr called the jump to another orbit a "quantum leap" or "quantum jump"— the electron never crosses the space in between. These words have now entered into our everyday language ("Cloning of Dolly the sheep was a quantum leap in genetic technology").

Bohr won the 1922 Nobel Prize for physics (his son Aage won the 1975 Nobel Prize for physics). In 1943 Bohr had to flee Nazi-occupied Denmark. First he went to England, and then to the US, where he worked at Los Alamos on the atomic bomb project. Before he left Denmark, Bohr dissolved his Nobel Prize medal in a bottle of acid and left it behind. After the war, he recovered the gold from the bottle and recast the medal.

See also Quantum Theory, p. 128.

1915 Theory of General Relativity

GERMANY · Albert Einstein (1879–1955)

Objects do not attract each other by exerting pull, but the presence of matter in space causes space to curve in such a manner that a gravitational field is set up. Gravity is the property of space itself.

This theory predicts that light should be bent by gravity and time should appear to run slower near a massive body like the Earth.

Einstein predicted that as a star's light passed the sun, it would be bent toward the Earth by the sun's gravity. Therefore, light from stars located behind the sun can be viewed during a total solar eclipse. The solar eclipse of May 29, 1919 provided an opportunity to test Einstein's theory. British scientists organized two expeditions to observe the solar eclipse: one to Principe Island, off the west coast of Africa and the other to Sobral in northern Brazil. When the news of scientists' results reached Einstein he wrote an excited postcard to his mother: "English expeditions have actually measured the deflection of starlight from the sun." Newton's law of gravitation (p. 38), which had ruled for more than two centuries, now faced a challenge.

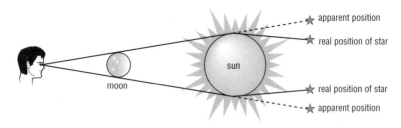

How a solar eclipse proved the theory of general relativity.

See also Special Theory of Relativity, p. 130.

1915

GERMANY

Wegener's Theory of Continental Drift

Alfred Wegener (1880–1930)

The Earth's land surface was once one big supercontinent. About 250 million years ago it broke up into the continents we know today, which have since drifted to their present positions.

The continents are still drifting across the surface of the Earth.

When Wegener published his theory in his book *The Origin of Continents and Oceans*, it was not taken seriously because of the lack of evidence. A famous scientist even pronounced it "utter, damned rot." This view prevailed until the 1960s, when new technologies revealed new geological and oceanographic evidence.

In 1962 the American geologist Harry Hammond Hess (1906–69) proposed his **seafloor spreading hypothesis**: the floor of the ocean is continuously being pulled apart along a narrow crack centered on a 39,000-mile-long ridge—known as the mid-ocean ridge—that threads its way down the middle of the North and South Atlantic and across the Pacific and Indian oceans. Volcanic mate-

rial rises from the Earth's mantle to fill the crack and continuously create new oceanic crust.

Now the **theory of plate tectonics** unifies the earlier ideas of continental drift and seafloor spreading. The theory says that the Earth's rigid outer shell, the lithosphere, consists of six major plates and several smaller ones which are in motion relative to each other. The plates are 43 to 93 miles thick and carry the continents and the oceanic basins on their backs like giant rafts. The plates are drifting like slow-moving ice floes over the mantle, the semifluid layer that underlies the lithosphere. Various plates are moving about ¾ inch per year (which is about the same speed as our nails grow).

Earthquakes and volcanoes are concentrated along the plate boundaries. The edges, or margins, of plates can move away from each other, push each other, or slide past each other. Colliding plates form mountains. If they move away from each other oceans are formed. Sliding plates form mid-oceanic ridges. The San Andreas Fault in California is a classic example of sliding plates.

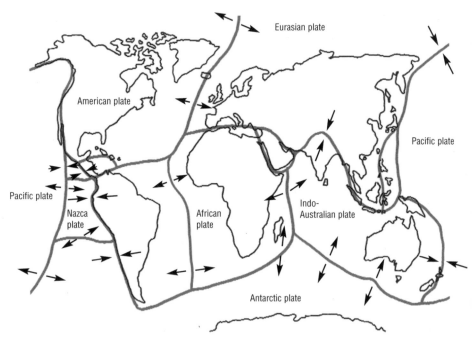

The Earth's rigid outer layer is divided into a number of plates.

The Brønsted–Lowry Concept of Acids and Bases

Johannes Brønsted (1879–1947)
Thomas Lowry (1874–1936)

An acid is a molecule or ion capable of donating a proton (that is, a hydrogen nucleus, H^+) in a chemical reaction, while a base is a molecule or ion capable of accepting one.

Stated simply, an acid is a proton donor and a base is a proton acceptor.

The Brønsted–Lowry concept extends Arrhenius' concept (p. 110) as it includes reactions that take place in the absence of water, such as reactions between ammonia and hydrochloric acid. In the same year that Brønsted and Lowry independently and concurrently arrived at their concept, the American chemist Gilbert Lewis (1875–1946) proposed a more generalized concept. According to the **Lewis concept of acids and bases**, an acid is a molecule or ion that can accept a pair of electrons; a base is a molecule or ion that can donate a pair of electrons. The Lewis concept also explains why metal oxides are basic (they contain the oxide ion, that donates two electrons) and non-metal oxides are acidic (the oxide ion accepts two electrons to share with the non-metal atom).

All chemistry students know that acids turn blue litmus paper red, and bases turn red litmus paper blue. Perhaps they do not know that the strongest known acid is an 80 percent solution of antimony pentafluoride in hydrofluoric acid, and the strongest known base is caesium hydroxide.

1924

FRANCE

De Broglie Waves

Louis de Broglie (1892–1987)

Like photons, particles such as electrons also show wave–particle duality, that is, they also behave like light waves.

If waves could act like particles, then why couldn't particles behave like waves? This reasoning helped de Broglie arrive at his theory, which played an important role in the development of wave mechanics.

The wavelength, λ (the lowercase Greek letter lambda), of the material particle, known as the "de Broglie wavelength," is given by the equation $\lambda = h/p$, where h is Planck's constant (p. 128) and p is the momentum (mass x velocity) of the particle. The wavelength of a photon and its momentum are also related in the same way.

The wave nature of electrons was proved experimentally in 1927. The experiment was similar to Young's double-slit experiment (p. 62), with one difference: the width of the slits was 0.1 nanometer. More recently, scientists have sent molecules as large as buckyballs (p. 192) through interference slits, creating characteristic interference patterns and proving that all atoms and molecules could behave like waves.

De Broglie was educated at the Sorbonne in Paris, where he presented his theory of the wave nature of electrons in his doctoral thesis. The examiners considered his revolutionary thesis too bizarre and asked Einstein to comment on it. Einstein replied: "It may look crazy, but it is really sound." The thesis was accepted, and five years later de Broglie was awarded the 1929 Nobel Prize for physics.

1924

Bose-Einstein Condensate

Albert Einstein (1879–1955)
Satyendra Nath Bose (1894–1974)

At temperatures close to absolute zero atoms and molecules lose their separate identities and merge into a single "super atom." This "super atom" is known as Bose-Einstein condensate.

Like solid, liquid, gas, and plasma (hot ionized gases found in fluorescent lights and the sun), Bose-Einstein condensate is a state of matter.

In quantum mechanics, elementary particles can, in some circumstances, behave like waves. The waves—which are really waves of probability—describe where a particle is most likely to be at a given moment. The uncertainty principle (p. 148) dictates that it is impossible to know the exact position of a particle. In 1924, while in Germany, Einstein predicted, based on ideas originally suggested by Indian-born Bose, that when atoms approach absolute zero (p. 92), the waves would expand and finally overlap; the elementary particles of which they are composed all merge into a single quantum state. This unique state is now known as Bose-Einstein condensate.

In 1995 American scientists were successful in creating Bose-Einstein condensate in a laboratory. They cooled rubidium atoms to a record 180 billionths of a degree below absolute zero (that is, −459.67°F). For up to 15 minutes at a time they not only turned the "bottle" that contained rubidium atoms into the coldest and stillest place in the universe, they also created a new form of matter.

1925 Pauli's Exclusion Principle

AUSTRIA Wolfgang Pauli (1900–58)

No two electrons in an atom can have the same quantum number.

A **quantum number describes certain properties of a particle, such as its charge and spin.**

One of the main applications of the principle is in the electron shell structure of atoms: an orbital or energy level cannot hold more than two electrons, one spinning clockwise, the other anticlockwise.

Electrons are grouped in shells, which contain orbitals. The shells are numbered ($n = 1$, 2, 3 etc.) outwards from the nucleus. These numbers are known as the "principal quantum numbers." An increase in n indicates an increase in energy associated with the shell, and an increase in the distance of the shell from the nucleus. The number of electrons allowed in a shell is $2n^2$. Each shell contains subshells or energy sublevels. A shell can only have n subshells. Each subshell is given a number and a letter (s, p, d, f, g, and so on). For example, the electron shell structure of lithium is $1s^2 2s^1$ (two electrons in "s" subshell of the first shell, and one electron in "s" subshell of the second shell; the superscript indicates the number of electrons in the shell).

The Pauli principle provided a theoretical basis for the modern periodic table (p.106). Pauli was awarded the 1945 Nobel Prize for physics for his principle.

See also Pauli's Neutrino Postulate, p. 154.

1926 The Schrödinger Equation

AUSTRIA   Erwin Schrödinger (1887–1961)

This complex mathematical equation describes the changing wave pattern of a particle such as an electron in an atom. The solution of the equation gives the probability of finding the particle at a particular place.

This fundamental equation in wave mechanics provides a mathematical description of the wavelike properties of particles.

Even if you do not know about the Schrödinger equation, you have probably heard about his cat. Well, not a real one, but a thought experiment referred to as **Schrödinger's cat**. Schrödinger, a scientific and mathematical genius, devised this experiment in 1935 to illustrate the probability of finding, say, an electron, at a particular place. He imagined a closed box containing a sample of radioactive material, a canister of cyanide, and a live cat. The ejection of a particle by the radioactive material provided a quantum event that would trigger the release of cyanide. After a certain time, if the quantum event had occurred the cat would have died; if not, the cat would still be alive. Schrödinger argued that the cat was neither dead nor alive until someone opened the box and observed it. But would the cat be alive or dead before opening the box? This paradox has not yet been fully resolved.

Schrödinger was awarded the 1933 Nobel Prize for physics for his work on wave mechanics.

1927

Heisenberg's Uncertainty Principle

Werner Heisenberg (1901–76)

It is impossible to determine exactly both the position and momentum of a particle (such as an electron) simultaneously.

The principle excludes the existence of a particle that is stationary. The uncertainty principle is a cornerstone of quantum theory (p. 128).

To measure both the position and momentum (momentum = mass x velocity) of a particle simultaneously requires two measurements: the act of performing the first measurement will "disturb" a particle and so create uncertainty in the second measurement. Thus the more accurately a position is known, the less accurately can the momentum be determined. The disturbance is so small that it can be ignored in the macroscopic (large-scale) world, but is quite dramatic for particles in the microscopic world. The uncertainty principle also applies to energy and time. A particle's kinetic energy cannot be measured with complete precision either. Heisenberg was awarded the 1932 Nobel Prize for physics for his discovery.

During World War II, Heisenberg reluctantly worked on the German atomic bomb project. In 1944 the American OSS (the CIA's wartime predecessor) sent an agent to attend a lecture by Heisenberg in Zurich in neutral Switzerland with the express instruction to shoot Heisenberg immediately if the lecture gave any hint that the German project was making progress. Luckily for science, Heisenberg did not mention the project during his lecture.

1928 Dirac's Antimatter Theory

ENGLAND　　　　　　　　　Paul Dirac (1902–84)

Every fundamental particle has an antiparticle—a mirror twin with the same mass but opposite charge.

The idea of antiparticles is now also applied to atoms—**antiatoms, which make up the antimatter.**

If you are a *Star Trek* fan you probably know that Starship Enterprise is powered by antimatter. Antimatter is not a stuff of science fiction; it does exist. In 1898 the British physicist Arthur Schuster (1851–1934) suggested the fascinating idea that an exotic type of matter could exist with properties that mirror those of ordinary matter: "If there is negative electricity, why not negative gold, as yellow as our own?" He added that this speculation was just "a dream." In 1928 Dirac, a gifted theoretical physicist, provided the mathematical basis for Schuster's dream. Dirac predicted that the electron, that is negatively charged, should have a positively charged counterpart: "This would be a new kind of particle, unknown to experimental physics, having the same mass and opposite charge as the electron. We may call such a particle an antielectron." Dirac's mathematical calculations also apply to other fundamental particles.

The discovery of antielectrons (now known as positrons, short for positively charged electron) in the cosmic radiation in 1932 by the American physicist Carl Anderson (1905–1991) vindicated Dirac's bold prediction. Twenty-three years later scientists at the University of California at Berkeley created the antiproton in a particle accelerator. When antimatter and ordinary matter meet, they annihilate each other and disappear in a violent explosion in which mass is converted into energy as dictated by Einstein's equation $E = mc^2$ (p. 132). The energy released in matter–antimatter annihilation is awesome: in a collision of protons and antiprotons the energy per particle is close to 200 times that available in a hydrogen bomb.

If matter and antimatter annihilate each other there is no likelihood of antimatter existing on the Earth, or even the solar system. Solar wind—the spray of charged particles emitted by the sun in all directions—would annihilate anti-matter. However, scientists speculate that antimatter could exist in other parts of the universe but so far they have found no evidence. This has not stopped them from creating antimatter in the laboratory. A team of scientists at CERN, the European nuclear physics lab in Geneva, did just that in 1995. For about 15 hours they fired a jet of xenon atoms across an antiproton beam. Collisions between antiprotons and xenon nuclei produced electrons and positrons. These positrons then combined with other antiprotons in the beam to make antihydrogen, the simplest antiatom.

If you find all this a bit too far-fetched, then what about the idea of an anti-universe—a universe parallel to ours? Enter it and you will find your antimatter counterpart: anti-you. Don't shake hands, you'll annihilate each other.

1929

AUSTRIA

Mach Number

Ernst Mach (1838–1916)

The ratio of the velocity of an object in air to the velocity of sound in air is termed the Mach number.

If the Mach number is 1, speed is called sonic. Below Mach 1 it's subsonic; above Mach 1, it's supersonic.

In 1929, Swiss engineer Jakob Ackeret (1898–1981) named the ration after the famous Austrian physicist Mach, who is best known for his work on airflow. However, his work on the philosophy of science greatly influenced Einstein.

The speed of sound in air is about 746 miles per hour. The denser the medium through which sound travels, the faster it travels. Therefore, the speed of sound varies at different altitudes, as at higher altitudes the air is less dense. When an airplane exceeds about 746 miles per hour, it produces a shock wave and breaks the so-called sound barrier. At Mach 2, the airplane will be flying at about 1491 miles per hour, that is, twice the speed of sound.

Captain Chuck Yeager, the legendary test pilot, was the first person to break the sound barrier, on October 14, 1947. His flight was in the Bell X-1 rocket under a US government research program. Tom Wolfe's famous book *The Right Stuff* (1979) and the 1983 movie of the same name vividly describe this flight.

1929

GERMANY

Berger's Experiments on Brain Waves

Hans Berger (1873–1941)

The brain generates electrical impulses, or waves, which can be physically recorded.

Until Berger's time, the internal workings of the brain were a complete mystery. Berger's experiments on brain waves have opened up the whole field of modern brain research.

In the 1890s, when Berger started working as a neuropsychologist, there was only one way to research into the brain—by dissecting it. But Berger thought differently: if the heart produces electrical signals (which can be recorded as an electrocardiogram), then the brain must also produce electrical signals. Berger had come up with a theory, but its translation into practice eluded him for nearly 30 years.

His persistence paid off when on July 6, 1924 he recorded electrical signals from a patient by attaching two electrodes to his scalp. He called the recording an electroencephalogram (EEG). Within years scientists in many countries repeated Berger's experiments with more sophisticated equipment and recorded different types of waves, now known as delta, theta, alpha, and beta waves. In spite of international acclaim, Berger was completely ignored in Germany. "The father of EEG" died a lonely, dejected man; his EEG lives on as a major medical diagnostic tool.

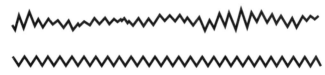

Berger's first EEG, showing two brain waves.

Hubble's Law

Edwin Hubble (1889–1953)

Galaxies are moving away from us and each other at an ever-increasing rate. The more distant the galaxy, the faster it is moving away from us.

This means that the universe is expanding like a balloon.

Hubble's law shows that the ratio of velocity of galaxies to their distances is a constant. This constant is known as the **Hubble constant**, which is the present rate of expansion of the universe—13.7 miles per second per million light years.

In the early 1920s Hubble began work at the Mount Wilson Observatory in California where he remained for the rest of his life. At Mount Wilson he used the 100-inch telescope, the largest of its day. With this telescope he discovered galaxies beyond our own galaxy, and classified them into three groups: elliptical, spiral, and irregular. We now know much more about galaxies: there are 125 billion (and still counting) galaxies in the presently observable universe and each contains billions of stars. Their diameters range from several thousand to one hundred thousand light years. But we can only know of galaxies that are within a certain radius, known as the **Hubble radius**, as galaxies larger than that radius will be traveling with the speed of light. The Hubble radius is estimated to be about 12 billion light years.

See also Big Bang Theory, p. 171; Olbers' Paradox, p. 70.

1931 Pauli's Neutrino Postulate

AUSTRIA   Wolfgang Pauli (1900–58)

The radioactive beta decay of an atomic nucleus in which a neutron turns into a proton and emits an electron does not seem to follow the law of conservation of energy (p. 88). To account for the "missing energy," Pauli postulated that a particle of zero charge and zero mass is released in such reactions.

A few years later Enrico Fermi **(p. 201) named the new particle a neutrino, which is Italian for "little neutral one."**

Neutrinos are no longer hypothetical particles. Their existence was confirmed in 1956. It is now believed that they also have a very small mass. Neutrinos are the most pervasive elementary particles in the universe. There are about 50 billion neutrinos for every electron, and they are everywhere. However, they cannot be seen and rarely interact with matter. Tens of thousands pass through our body every second. There are three known types of neutrino—muon, tau, and electron—and they are all created in the center of the sun, supernovas, and the cosmic rays hitting the upper atmosphere.

Neutrinos' ghostly behavior has prompted the American writer John Updike (known for his Rabbit series of novels) to write a poem, entitled "Cosmic Gall." An excerpt:

At night, night they enter at Nepal
And pierce the lover and his lass
From underneath the bed—you call
It wonderful; I call it crass.

1931

USA

Pauling's Theory of the Chemical Bond

Linus Pauling (1901–94)

A framework for understanding the electronic and geometric structure of molecules and crystals. An important aspect of this framework is the concept of hybridization: in order to create stronger bonds, atoms change the shape of their orbitals (the space around a nucleus in which an electron is most likely to be found) into petal shapes, which allow more effective overlapping of orbitals.

A chemical bond is a strong force of attraction linking atoms in a molecule or crystal. Pauling was the first to use quantum mechanics to explain chemical bonds. His theory is a milestone in the development of modern chemistry.

Pauling was awarded the 1954 Nobel Prize for chemistry for his work on chemical bonds. After World War II Pauling worked energetically to awaken the conscience of society to its new responsibilities in a nuclear age. In 1962 he was awarded the Nobel Peace Prize.

He is the only person ever to receive two unshared Nobel Prizes. Pauling was also the champion of another cause: large doses of vitamin C are effective in the prevention of the common cold and flu. This theory, however, has not won any significant scientific support.

Pauling once said: "The best way to have a good idea is to have lot of ideas." Most chemistry students are familiar with another of Pauling's good ideas: the **electronegativity scale**, which ranks elements in order of their electronegativity (0.7 for cesium and francium to 4.0 for fluorine).

1931

Dirac's Conception of the Magnetic Monopole

Paul Dirac (1902–84)

The magnetic monopole is a hypothetical particle that carries a basic magnetic charge—in effect, a single north or south magnetic pole acting as a free particle.

A magnetic monopole is analogous to electric charge.

Ever since Dirac ("There is no God and Dirac is his prophet," Wolfgang Pauli, p. 146, used to say about the great physicist) predicted its existence, the magnetic monopole has intrigued physicists. A magnet broken in two does not form two monopole magnets, one with a north pole only and the other with a south pole only; it breaks into two similar dipole magnets. Even if this process of breaking a magnet in two is carried to the smallest particle, it ends up with a particle with two poles—north and south.

After formulating his antimatter theory (p. 149) Dirac tried to link the electric and magnetic phenomena by predicting the existence of the monopole. He said that even if one monopole existed somewhere in the universe, it would explain why electric charge comes only in multiples of the charge on the electron. No one has observed a monopole yet. "If you want a Nobel Prize the hard way, you know what to look for," was the advice the late Brian L. Silver gave to budding scientists in his book *The Ascent of Science* (1998).

1931

AUSTRIA

Gödel's Incompleteness Theorem

Kurt Gödel (1906–78)

Every consistent theory must contain propositions that can be neither proved nor disproved according to its own defining set of rules.

The theorem proved the "incompleteness of mathematics." Its implication is that all logical systems of any complexity are incomplete.

Gödel published his theorem—one of the most extraordinary achievements in mathematics in the twentieth century—in 1931, a year after taking his doctoral degree from the University of Vienna. When the Nazis occupied Austria in 1938, he emigrated to the US. He held a chair at the Institute of Advanced Studies at Princeton from 1953 until his death. Einstein, also at Princeton during that time, was Gödel's closest friend.

In the later years of his life Gödel's behavior became increasingly eccentric. He withdrew from all human contact and received communications only through a crack in the door of his office. He became notorious for wearing ski masks with eye holes wherever he went. Toward the end of his life he feared that he was being poisoned and stopped eating altogether. He died sitting in a chair in a hospital room in Princeton.

He once said: "Either mathematics is too big for the human mind or the human mind is more than a machine." Some people have extended Gödel's theorem to claim that it will be impossible to create a machine as intelligent as a human being.

The Chandrasekhar Limit

Subrahmanyan Chandrasekhar (1910–95)

The maximum possible mass of a white dwarf star is 1.44 times the sun's mass.

The Chandrasekhar limit is a physical constant. A star of greater mass will become a neutron star or a black hole, because of the force of gravity.

The sun has lived 4600 million years as a stable star and many billion years lie ahead. Once it has completely exhausted its nuclear fuel, it will shrink into a "white dwarf"—no larger than the Earth but so heavy that a teaspoonful of its matter would weigh thousands of pounds. The white dwarf is so hot that it shines white-hot. No white dwarfs have been found with a mass greater than 1.44 times the sun's. This limiting solar mass was predicted by Chandrasekhar, an Indian-born astro-physicist who was awarded the 1983 Nobel Prize for physics for his work on the structure and evolution of stars.

A massive star does not evolve into a white dwarf: it explodes as a supernova, which ejects an enormous amount of matter and even outshines the entire galaxy for a few days. The remaining matter forms a neutron star, only a few miles across, which contains tightly packed neutrons. These neutron stars do not glow and are so heavy that even a pinhead of their matter would have a mass of a million tons.

Sometimes the crushing weight of a dying star like a neutron star squeezes it into a point with infinite density. At this point, known as singularity, mass has no volume and both space and time stop. The singularity is surrounded by an imaginary surface known as the event horizon, a kind of one-way spherical boundary. Nothing—not even light—can escape the event horizon. Matter falling into it is swallowed and disappears forever. That's why scientists call these regions of space–time black holes. If an astronaut were to pass through the event horizon of a black hole, gravita-tional forces would stretch his or her body in the shape of very long spaghetti, and when this very dead spaghetti slams into the singularity of the black hole, the astronaut's remains would be ripped apart into atoms.

The radius of a black hole is the radius of the event horizon surrounding it. This is called the **Schwarzschild radius**, after the German astronomer Karl Schwarzschild (1873–1916) who in 1916 predicted the existence of black holes. The Schwarzschild radius is roughly equal to three times the weight of the black hole (in solar masses). A black hole weighing as much as the sun would have a radius of 1.9 miles; one with the mass of Earth would have a radius of only ⅕ inch; and one with the mass of a small asteroid would be roughly the size of an atomic nucleus. A black hole's weird effects occur within ten Schwarzschild radii of its center. Beyond this rather limited distance, the only effect is through the black hole's normal gravitational pull. So, contrary to popular belief, a black hole is not like a cosmic vacuum cleaner that sucks in everything around it.

See also Hawking's Black Hole Theory, p. 188.

1934

Oliphant's Concept of Isotopes of Hydrogen

Marcus Oliphant (1901–2000)

Hydrogen has three isotopes: hydrogen-1 (ordinary hydrogen: one neutron), hydrogen-2 (deuterium: one proton, one neutron), and hydrogen-3 (tritium: two protons, one neutron).

Oliphant was the first to create tritium. This discovery was fundamental to the development of the atomic bomb.

In the 1930s Oliphant, an Australian-born nuclear physicist, worked on splitting the atom at the Cavendish Laboratory in Cambridge with Ernest Rutherford (p. 136). By that time a heavy form of hydrogen called deuterium had been discovered. Oliphant bombarded deuterium nuclei with other deuterium nuclei and produced a new isotope, tritium. In 1943 he moved to the US, where he perfected the technique of purifying uranium-235, which was used to produce the first atomic bomb.

In 1950 Oliphant returned to Australia, where he served as the Governor of South Australia from 1971 to 1976. He died at the age of 98. Among the numerous obituaries published around the world, the one published by the journal *Current Science* (October 10, 2000) started with an anecdote: "Popular radio comedian Fred Allen once asked an actor impersonating a physicist as to why anyone would spend his time trying to smash atoms. The reply, delivered in a thick German accent, 'Vell, someday someone might vant half an atom' brought the house down." Oliphant, a man with an informal personality, zest for life, and boisterous laughter, would have loved the joke.

1935

US

The Richter Scale

Charles Richter (1900–85)

A scale ranging from 0 to 9 to measure the magnitude of earthquakes.

Earthquakes of magnitudes 3.5 and less are usually not felt. Earthquakes of magnitudes between 5.5 and 6.0 cause slight damage to buildings. Severe earthquakes have magnitudes greater than 7.0.

It's a popular misconception that the Richter scale is actually a piece of equipment, like a bathroom scale. Seismologists have complained that journalists ask them for a picture of the Richter scale. It's a bit like asking for a picture of a mile, they say. Ironically, Richter, a US seismologist, developed his scale because he was tired of journalists in earthquake-plagued southern California asking him about the relative size of earthquakes.

An earthquake generates a series of shock waves, known as seismic waves. The Richter scale is simply a numerical scale which gives the magnitude of an earthquake by calculating the energy of shock waves at a standard distance. The scale is logarithmic, so each additional point represents a tenfold increase in severity. Thus a magnitude 7.0 earthquake is 10 times as powerful as one of magnitude 6.0 and 100 times as powerful as one of magnitude 5.0. In terms of energy, magnitude units rise even faster: one unit represents an increase in the energy of roughly 33 times. A magnitude 7.0 earthquake, for instance, unleashes about 1000 times the energy released by a magnitude 5.0 earthquake.

1936

RUSSIA

Oparin's Theory of the Origin of Life

Aleksandr Oparin (1894–1980)

In the Earth's early atmosphere simple inorganic compounds combined to form complex organic compounds, which formed the first living cell.

Recent advances in molecular biology have put this theory on trial.

Oparin suggested that early in the Earth's history the atmosphere was rich in hydrogen. Simple inorganic hydrogen compounds such as water, methane, and ammonia could form organic compounds. Gradually these organic compounds fell from the atmosphere to the ground, where the rain—which occurred when the Earth cooled and water vapor condensed—washed them into pools and ultimately into oceans. Over millions of years the organic molecules in this "primeval soup" joined together into long chains of proteins and DNA molecules until a cell appeared which possessed the right kind of reactions and right kind of compounds to be considered an organism. This first cell could replicate itself and therefore it filled the bill for the first living organism.

In 1953 Stanley Miller, a student at the University of Chicago, provided the first experimental support to Oparin's theory. He subjected a mixture of methane, ammonia, water vapor, and hydrogen to a series of electrical charges. He imagined this to be a rough duplication of conditions on the primitive Earth when the primeval soup was subjected to bolts of lightning. After a week, the inorganic molecules had joined to form amino acids, the building blocks of life.

1937   The Turing Machine

ENGLAND   Alan Turing (1912–54)

A theoretical computer with two or more possible states, which can react to an input to produce an output.

The Turing machine was a major landmark in the development of digital computers.

A Turing machine contains a tape of infinite length divided into cells, each inscribed with 0 or 1. A read–write head can read or write to a cell in its current location, and also move step by step either way along the tape. Turing proposed this imaginary machine to give a mathematically precise definition of "algorithm." The machine obeyed instructions set out in an algorithm.

During World War II, Turing built a computing device which was used to break the secret communication code, called Enigma, used by the Germans. In 1950 he suggested that it must be possible to program computers to acquire human intelligence and devised a test that will verify computer intelligence. Turing suggested that if the response from the computer was indistinguishable from that of a human, the computer could be said to be intelligent. The **Turing test** is used today to determine if a computer can really imitate human intelligence. A computer and a person are interrogated by text messages. If the interrogator could not distinguish which answer came from the person and which from the computer, then the computer could be called intelligent.

Game Theory

John von Neumann (1903–57), Oskar Morgenstern (1902–77),
John Nash (b. 1928)

**A mathematical method of analyzing strategic behavior—how people
behave when placed in competitive situations.**

The theory has applications in economics, computer science, psychology, sociology, politics, warfare, evolution, the stock market, and many other fields.

According to the game theory all games have three things in common: rules, strategies, and payoffs. Games include zero-sum games (each player benefits at the expense of others), non-zero-sum games, cooperative games (people can make bargains), and games of complete information. The equilibrium of a game is called the **Nash equilibrium**, a solution that maximizes everyone's benefit.

The game theory was born when von Neumann realized that poker, a game he played occasionally was not guided by probability theory (p. 32) alone, and that "bluffing," a strategy to hide information from other players, was also crucial. The theory was extended by Neumann and Morgenstern in 1944, and Nash in 1949. Nash wrote his seminal paper, "Non-cooperative Games," while studying for his PhD at Princeton University. A few years later he was diagnosed with paranoid schizophrenia. In the early 1990s he overcame his disease and resumed his work. He was awarded the 1994 Nobel Prize for economics. Sylvia Nasser's book *A Beautiful Mind* (1998) and the movie of the same name (2001) present Nash's dramatic life story. (Von Neumann died before the economics Nobel was introduced.)

1938 Nuclear Fission

GERMANY   Otto Hahn (Germany 1879–1968), Lise Meitner (Austria 1878–1968)
Fritz Strassmann (Germany 1902–1980)

The breaking up of the nucleus of a heavy atom into two or more lighter atoms. Energy is released during nuclear fission.

Nuclear fission takes place in atomic reactions such as in an atomic bomb.

In 1938 Hahn, working with Strassmann, made an amazing discovery—uranium nuclei bombarded with slow-moving neutrons gave barium—but he was unable to explain it. "Perhaps you can come up with some sort of fantastic explanation," Hahn wrote to Meitner, a long-time associate. (Together, in 1917, they had discovered a new element, protactinium, but now Meitner was in exile in Sweden.) Within a few days, Meitner showed that the uranium nucleus, after absorbing a neutron, had split into two roughly equal pieces, barium and krypton. She also calculated that the process would release much energy. Her nephew, Otto Frisch (1904–79), also a noted physicist, named the effect "nuclear fission" because of its resemblance to biological "fission" (cell division).

Hahn published the chemical evidence for nuclear fission without listing Meitner as a coauthor. In 1944 he was awarded the Nobel Prize for chemistry for his discovery. She did not get a share in the Nobel Prize. However, she was awarded the 1966 Enrico Fermi Prize for her work on nuclear physics. In 1997 the element 109 was named meitnerium (Mt) in her honor.

1938

US

Bethe's Theory of Energy Production in Stars

Hans Bethe (b. 1906)

Energy in stars is produced by hydrogen fusion reactions.

In nuclear fusion the nuclei of light atoms combine at very high temperatures and release enormous amounts of energy that is radiated from the surface of the star as heat and light.

"Twinkle, twinkle little star/How I wonder what you are". Perhaps you no longer retain the childhood fascination for the stars in the sky, but stars still twinkle in the eyes of most astrophysicists. One of them formulated the first detailed theory for the formation of energy by stars.

An ordinary star is one of the simplest entities in nature: it is a sphere of gas that is by mass 73 percent hydrogen, 25 percent helium and 2 percent other elements. The temperature in the center of a star is very high—high enough to fuse four nuclei of hydrogen together to form one helium nucleus. This process, which generates enormous amounts of energy, is sometimes known as the carbon–nitrogen–oxygen (CNO) cycle. These elements act as catalysts and they are not consumed in the reaction.

Bethe's theory led other scientists to develop hydrogen bombs, which are much more destructive than atom bombs. Bethe, who was born in Germany and moved to the US in 1937, was awarded the 1967 Nobel Prize for physics for his theory.

See also The Chandrasekhar Limit, p. 158.

1940 Transuranium Elements

US

Edwin McMillan (1907–91)
Glenn Seaborg (1912–99)

Elements heavier than uranium in the periodic table (transuranium elements) are made artificially. Uranium (U, atomic number 92) is the heaviest element known to exist naturally in detectable amounts on the Earth.

More than 20 transuranium elements have been made since 1940.

Enrico Fermi (p. 201) set the stage for creating new elements when in 1933 he showed that the nucleus of most elements would absorb a neutron, changing the element into a new atom. It occurred to him that it should be possible to prepare new atoms by bombarding uranium nuclei with free neutrons. He tried the experiment in 1934 but failed. In 1940 McMillan, a nuclear physicist, produced and identified the first artificial element, neptunium (Np, 93). In 1940 Seaborg, a chemist, succeeded in creating element 94, which was named plutonium (Pu).

McMillan and Seaborg were awarded the 1951 Nobel Prize for chemistry for their work.

Nine transuranium elements have been named after scientists: curium (Cm, 96: Marie and Pierre Curie, p. 126), einsteinium (Es, 99: Albert Einstein, p. 130), fermium (Fm, 100: Enrico Fermi, p. 201), mendelevium (Md, 101: Dmitri Mendeleev, p. 106), nobelium (No, 102: Swedish chemist Alfred Nobel, 1833–96, known for his bequest for the foundation of the Nobel Prizes), rutherfordium (Rf, 104: Ernest Rutherford, p. 136), seaborgium (Sg, 106: Glenn Seaborg), bohrium (Bh, 107: Niels Bohr, p. 139) and meitnerium (Mt, 109: Lise Meitner, p. 165).

1940

US

Asimov's Three Laws of Robotics

Isaac Asimov (1920–92)

First Law: **A robot may not injure a human being or, through inaction, allow a human being to come to harm.**
Second Law: **A robot must obey orders given it by human beings, except where such orders would conflict with the First Law.**
Third Law: **A robot must protect its own existence as long as such protection does not conflict with the First or Second Law.**

From *Handbook of Robotics*, 56th Edition, 2058 AD (as quoted in *Runaround*, a science fiction short story by Asimov). Isaac Asimov's laws are not laws of science. They are nothing but fiction, but even today researchers in artificial intelligence would like their intelligent machines to follow these laws.

The word robot was introduced in the English language from a 1921 play *RUR* (*Rossum's Universal Robots*) by Czech playwright Karel Çapek. Rossum, a fictional Englishman, used biological methods to mass produce robots to serve humans. Asimov started writing science fiction stories about robots at the age of 20. Asimov's fictional robots were machines designed by engineers, and they were built with three laws implanted in their "positronic brains." Russian-born Asimov—he arrived in New York when he was three—was a prodigious science and science fiction writer. In his 52-year writing career he published nearly 500 books. His most famous science fiction works are the Foundation series books, which chronicle the history of a Galactic Empire.

1946

Radiocarbon Dating

Willard Libby (1908–80)

The radioactive isotope of carbon, carbon-14, is present in all living things. When life stops carbon-14 begins to decay. From the rate of decay, the age (or time of death) of an organism can be calculated.

Radiocarbon dating can be used to estimate the age of any organic material.

Each radioactive isotope decays (breaks down) at a rate known as its half-life. For example, the half-life of carbon-14 is 5730 years. This means that a sample of 1000 carbon-14 atoms would contain only 500 atoms after 5730 years (the other 500 having turned into stable nitrogen-14). After another 5730, another 250 atoms would decay into nitrogen-14, and so on. Living things go on absorbing carbon-12 and carbon-14 until the time of their death. Once an organism dies, carbon-14 begins to decay. As a result the ratio of carbon-12 to carbon-14 changes with time. By measuring this ratio, it can be determined when the organism died.

Geologists and anthropologists now used radiocarbon dating to measure accurately the age of very old wood, bones, fossils, and other artefacts. The technique has helped us to understand the history of the Earth and the development of life.

Libby, a chemist who worked on the first atomic bomb project, was awarded the 1960 Nobel Prize for chemistry for his work. He was a tireless advocate of the peaceful uses of nuclear technology.

Wiener's Cybernetics

Norbert Wiener (1894–1964)

Cybernetics is the study of control and communication in both machines and animals.

The term "cybernetics" is derived from the Greek word for helmsman, kybernetes. It now refers to the theory of computer control systems involved in automation, with emphasis on comparing machines with the nervous system of humans.

Wiener, a brilliant mathematician, published his ideas on cybernetics in his book *Cybernetics: Control and Communication in the Animal and the Machine (*1948). In this he explained that "cybernetics attempts to find the common elements in the functioning of automatic machines and of the human nervous system." This knowledge, he believed, would improve the efficiency of machines. The book also introduced terms such as "input," "output" and "feedback." Feedback—controlling the performance of a system by using output to modify its input without human intervention—is an important concept in cybernetics. For example, a thermostat in a gas space heater works properly because it senses the temperature (a measure of output) of the system and feeds it back to the system to modify gas supply (input).

At the Massachusetts Institute of Technology, where he worked for most of his life, Wiener was more famous for his eccentricities and unworldliness than for his outstanding theories. He once asked a small girl in the street whether she might be able to direct him towards Brattle Street. The girl giggled: "Yes, daddy," she said, "I'll take you home."

1948

US

The Big Bang Theory

George Gamow (1904–68)

The universe began when a single point of infinitely dense and infinitely hot matter exploded spontaneously. The debris of this explosion began to fly away from the explosion point and is still flying and will keep on flying indefinitely. All the galaxies, stars, and planets were formed from this debris.

Time begins at the Big Bang, which happened about 12 billion years ago.

In 1927 the Belgian astronomer Georges Lemaître (1894–1966) suggested that at some time in the remote past all the matter in the universe was concentrated at one point. The universe began when this "primeval atom" exploded. This idea was further developed by Gamow, who showed that as the universe began from a "fireball;" leftover warmth from this primeval fireball still filled the universe. This leftover radiation should now have a temperature of 3 Kelvin, or −454°F. This radiation was detected in 1965 and, as predicted by Gamow, it has a temperature of −454°F.

Will the universe expand forever? There are two opposing views: the expansion may continue forever, or some day it may collapse back into the "primeval atom." It is known as the Big Crunch. The name Big Bang was given by Fred Hoyle, who believed in the opposing steady state theory (p. 172). It was meant to be a put-down when he first used it scornfully in a radio talk in 1950.

1948 Steady State Theory

ENGLAND

Herman Bondi (UK, b. 1919) Thomas Gold (US, 1920–2004)
Fred Hoyle (UK, 1915–2001)

The universe has no beginning and will have no end. It is constantly producing matter and is expanding.

This theory is now considered flawed and the big bang theory is widely accepted.

The steady state theory includes the idea of spontaneous creation of matter. On the other hand, the big bang theory (p. 171) assumes that all matter that now exists also existed in the past. New matter is not being created. The steady-state theory agrees with the big bang theory on one point: the universe is expanding.

The big bang theory holds that the universe had a beginning and will someday have an end. "The old problem of the beginning and end of the universe does not arise at all in the steady state theory, for the universe did not have a beginning and will not have an end," according to Fred Hoyle. "Every cluster of galaxies, every star, every atom had a beginning but the universe itself did not."

Observational and experimental data favor the big bang theory and it is now considered the standard theory of the origin, structure, and future of the universe. Hoyle was a staunch supporter of the steady state theory and never gave up his belief in it.

See also Hoyle's Theory of the Origin of Elements, p. 179.

Murphy's Law

Edward A. Murphy Jr (b. 1917)

If anything can go wrong, it will.

Murphy's law is expressed in various humorous axioms stating that anything that can possibly go wrong will go wrong. In mathematical form it is expressed as 1 + 1 * 2, where * stands for "hardly ever."

At Edwards Air Force Base in California, John Paul Stapp and George E. Nichols were working on an aerospace project designed to test how much sudden deceleration a person can stand in a crash. Murphy came from another laboratory bringing a set of gauges that were supposed to measure the deceleration more accurately. However, the gauges measured no deceleration at all because they were wired wrongly.

Thoroughly annoyed, Murphy cursed the technician responsible and muttered something approximating his immortal law. Murphy's law was born. After 50 years, in 1999, Stapp, Nichols, and Murphy were awarded the Ig Noble Prize, a spoof on the Nobel Prizes which is awarded annually by the science humor magazine *Annals of Improbable Research* to honor achievements that "cannot or should not be reproduced."

Murphy's law is not a scientific law, nor is the **Peter principle** (after Laurence Peter, 1919–90): Employees within an organization will advance to their highest level of competence and then be promoted to and remain at a level at which they are incompetent.

See also Tumbling Toast Theory, p. 196.

1950 Oort Cloud of Comets

HOLLAND   Jan Oort (1900–92)

The solar system is surrounded by a cloud of billions of comets.

Oort's hypothesis is now widely accepted. **The halo-like cloud, which lies far beyond the orbit of Pluto, is known as the Oort cloud.**

There are about two to five trillion comets that circle the solar system in the Oort cloud between 20,000 and 100,000 astronomical units from the sun (one astronomical unit is the distance between the Earth and sun, about 93 million miles). Comets in the Oort cloud are not packed like sardines—neighboring comets are typically tens of millions of miles apart. The Oort cloud is sometimes called the Siberia of comets because of its freezing temperatures, as low as −454°F. Occasionally a comet is deflected into the orbits of inner planets by the gravitational pull of nearby stars.

Oort, a distinguished astronomer, also determined the structure, size, mass, and motion of the Milky Way.

In 1951 the American astronomer Gerard P. Kuiper (1905–73) suggested that there is another reservoir of comets—now known as the **Kuiper belt**. The Kuiper belt extends between 35 and a few hundred astronomical units from the sun, beyond the orbit of Neptune. It is a like a CD: the hole covers the solar system inside Neptune.

See also Whipple's Theory of Comets, p. 175.

1950

US

Whipple's Theory of Comets

Fred Lawrence Whipple (1906–2004)

A typical comet has three parts: a frozen central part called the nucleus, a fuzzy cloud surrounding the nucleus called the coma (or head), and a tail consisting of gas and dust. The nucleus, usually only a few miles across, is a "dirty snowball" made of grains of frozen mass of water, methane, ethane, carbon dioxide, ammonia, and many other gases.

Before Whipple proposed his theory astronomers believed that comets consisted of one or a few large stony rocks or even a "sandbag" of small particles. The size of nuclei was also overestimated to be several hundred miles.

Whipple not only coined the evocative phrase "dirty snowball," but he also came up with an equally evocative idea: a comet is like a jet engine. Like the heated gases erupting from a jet engine, the evaporating gases from the nucleus exert a force on the nucleus. This force gives the comet its independent thrust. "When I first realized the jet action of comets," 79-year-old Whipple told *Time* magazine in 1985, "Boy! That was a thrill."

In 1986 the European Space Agency's *Giotto* spacecraft proved that Whipple's theory was fairly accurate when it took close-up photographs (from a distance of 298 miles) of the nucleus of Halley's comet: the comet's nucleus resembled a fluffy snowball coated with a crust of black material and spouting jets of vaporized ice.

See also Oort Cloud of Comets, p. 174.

1953 DNA's Double Helix Structure

ENGLAND

Francis Crick (US, 1916–2004)
James Watson (UK, b. 1928)

The self-reproducing genetic molecule DNA has the form of a double helix.

The structure explained how the DNA stores information and replicates itself. The discovery of DNA's "code of life" revolutionized molecular biology and advanced our understanding of life.

The helical strands of DNA (deoxyribonucleic acid) consist of chains of alternating sugar and phosphate groups. Four types of base—adenine (A), cytosine (C), guanine (G), and thymine (T)—form the rungs of the DNA ladder, which can only be linked by hydrogen bonds in four combinations: A–T, C–G, T–A, G–C. Life's code is based on the order of these four bases, which DNA carries from one generation to the next. The sequence of base pairs along the length of the strands is not the same in DNAs of different organisms. It is this difference in the sequence that makes one gene different from another.

Crick and Watson shared the 1962 Nobel Prize for physiology or medicine with British scientist Maurice Wilkins (1916–2004) for their discovery.

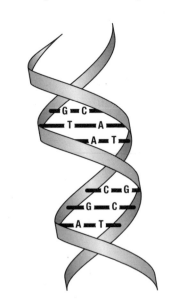

A section of DNA.

1956

Hodgkin's Structure of Biological Molecules

Dorothy Crowfoot Hodgkin (1910–94)

The structure of large organic molecules can be determined by x-ray analysis.

Hodgkin discovered the structure of vitamin B12 and insulin. Once chemists know the structure of a molecule, they can synthesize it.

X-ray crystallography (see Bragg's Law, p. 138) is used to study the structure of molecules by shining x-rays through them. The molecules cause the x-rays to split into a complex pattern which can be analyzed mathematically. x-ray crystallography works fine for simple molecules but the task becomes extremely complex when studying large organic molecules. Hodgkin pioneered the use of computers in x-ray crystallography and in 1956 produced a three-dimensional picture of vitamin B12 (which has nearly 100 atoms). Hodgkin

was awarded the 1956 Nobel Prize for chemistry for her work. In 1969 she determined the structure of insulin, a task that took her 34 years (the B12 task took her six long years).

Hodgkin was born in Cairo, where her father was working at the time. She started her secondary education at eleven when her parents moved to England. "So when I first went to secondary school, I was rather behind, if anything," she once recalled in a radio interview. "I was *terribly* behind in arithmetic." But she worked hard and by the end of the year was first in the class. And then she also became the first to unravel the complex world of large biological molecules.

Lee and Yang's
Concept of Parity

Tsung-Dao Lee (b. 1926)
Chen Ning Yang (b. 1922)

Parity is not conserved in weak interactions between elementary particles.

Elementary particles interact by four types of forces: gravity (the attraction between all matter), electromagnetism (the force between charged particles), the strong force (it holds together an atomic nucleus), and the weak force (a type of nuclear force).

The existence of antimatter (p. 149) leads to the idea of symmetry: that is, every particle has a mirrorlike twin. An antiparticle would look just like the ordinary particle, except that left would be switched with right. Physicists call it reversing the parity (parity is just another word for left–right or mirror symmetry).

The law of conservation of parity says that the laws of physics are identical in right- or left-handed systems of coordinates. But nature's symmetry is flawed. Certain interactions of elementary particles always produce a particle always spinning in the same direction. For example, when an atom emits a neutrino it always spins in the same direction— left-handedly. As many elementary particles display a preference for left over right, the universe seems to be left-handed. Why? Physicists do not know.

In 1956 Lee and Yang suggested that the evidence for left–right symmetry was weak in interactions involving the weak force. This prediction was soon confirmed experimentally by other physicists. The discovery of asymmetry won Chinese-born Yang and Lee the Nobel Prize for physics just a year later.

1957

Hoyle's Theory of the Origin of Elements

Fred Hoyle (1915–2001)

Most of the elements heavier than hydrogen in the universe are created, or synthesized, in stars when lighter nuclei fuse to make heavier nuclei. The process is called "stellar nucleosynthesis."

The theory explains how chemical elements are manufactured inside stars.

Our sun burns, or fuses, hydrogen to helium. This process occurs during most of every star's lifetime. After a star exhausts its supply of hydrogen, the star burns helium to form beryllium, carbon, and oxygen. When the star exhausts its supply of helium, it shrinks and its temperature rises to 1000 million degrees. The rising temperature triggers a new series of reactions in which carbon, oxygen, and other elements combine to form iron and nickel. When the star has burned everything into iron and nickel, it explodes as a supernova. The elements heavier than nickel are formed during supernova explosions.

This theory was proposed by Hoyle (with assistance from W. A. Fowler and husband and wife team Geoffrey and Margaret Burbidge) in his monumental paper in 1957. Hoyle, the eminent astronomer, also proposed, with Thomas Gold and Herman Bondi, the steady-state theory of the origin of the universe in 1948 (p. 172). He also coined the term "big bang theory" (p. 171) to distinguish it, derisively, from his steady state theory. He was also a popular science fiction novelist (*Black Cloud*, 1957).

Calvin Cycle in Photosynthesis

Melvin Calvin (1911–97)

The cycle of chemical reactions by which plants turn carbon dioxide and water into sugar during photosynthesis.

The cycle provided an insight into the nature of photosynthesis. This knowledge has promoted interest in artificial photosynthesis to harness solar energy (p. 51).

In the Calvin cycle, which takes place in most plants, the initial product of the dark reactions (reactions in which carbon is converted into sugar) is a compound with three carbon atoms per molecule. Hence these plants are known as C3 plants.

A small group of plants—including maize, sorghum, and sugar cane—are formed by a different cycle, known as the **Hatch and Slack pathway**. In this cycle the initial product of the dark reactions is a compound with four carbon atoms per molecule. These plants, known as C4 plants, can assimilate carbon dioxide at about twice or more the rate of C3 plants and hence grow faster. The Hatch and Slack pathway was discovered in 1966 by two Australian scientists, M. D. Hatch and C. R. Slack.

Calvin, a biochemist, introduced carbon dioxide made from radioactive carbon-14 in plants to study the photosynthesis cycle. His idea of tracing radiation from carbon-14 is now used widely in biochemistry. Calvin was awarded the 1961 Nobel Prize for chemistry for his ground breaking work.

1959

US &
ENGLAND

Chemical Structure
of Antibodies

Gerald Edelman (US, b. 1929)
Rodney Porter (UK, 1917–85)

Antibody molecules are in the shape of the letter Y, with a stem and two angled branches. Each branch is composed of one light chain and one half of a heavy chain in side-by-side arrangement. The stem is made up of the remaining halves of the heavy chains.

The angled branches bind to foreign antigens to destroy them.

Antibodies (also called immunoglobulins) are an important part of our immune system's response to infections. They are specialized protein molecules produced by white blood cells. There are five classes of antibodies: IgA (immunoglobulin A), IgD, IgE, IgG, and IgM. There are probably millions of different antibodies in the blood, each of them reactive against one particular antigen (foreign substances such as bacteria, viruses, or toxins).

In 1959 Edelman and Porter separately and independently published their fundamental studies of the molecular structure of antibodies. Brought together, their studies complemented each other. Edelman and Porter were awarded the 1972 Nobel Prize for physiology or medicine. Many novel and fascinating aspects on problems in molecular biology and genetics have grown out of their research, said their Nobel Prize citation. "We have now a new and firmer grasp of the question of the role of immunity as defence against and as cause of disease. Our possibilities to make use of immune reactions for diagnostic and therapeutic purposes have improved."

1961 Horsfall's Theory of Cancer

US

Frank Horsfall (1906–71)

Cancer is caused by changes in the DNA of cells.

The theory provided a basis for research on cancer.

Cancer is any one of a group of more than 100 different and distinctive diseases in which some body cells grow uncontrollably. It is the result of one of the body's billions of cells being damaged in such a way that it replicates abnormally to produce enormous numbers of identically damaged cells. These cells in turn replicate to form what is known as a cancer, or a tumor. As the behavior of a cell is controlled by its DNA, cancer is caused by damage to DNA. Damaged or mutated DNA may be inherited or damage may be caused by carcinogens. A carcinogen is any agent that causes cancer; for example, radioactivity, high doses of x-rays and of ultraviolet rays, and certain chemicals.

Horsfall, a brilliant clinician and virologist, was the director of the Sloan-Kettering Institute for Cancer Research in New York when he made his discovery that cancer resulted from changes produced in the genetic machinery (DNA) of the cell. He said that viral and chemical carcinogens are inter-related. This unifying concept of cancer marked an important shift in the study of cancer and resulted in rapid advances in our knowledge of cancer. The man who gave so much to cancer research died of cancer at the age of 65.

1961

The Drake Equation

Frank Drake (b. 1930)

The number of advanced technical civilisations in our galaxy can be predicted by a simple equation.

Are you there, ET?

In its simplest form, the equation works as follows. To find out the number (N) of advanced technical civilizations in the Milky Way, we need to know:

- how many stars are born each year in our galaxy (R)
- how many of these stars have planets (p)
- how many of these planets are suitable for life (e)
- on how many planets life actually appears (l)
- on how many planets life evolves to an intelligent form (i)
- on how many planets the intelligent life can communicate to other worlds (c)
- the average life of these advanced civilizations (L).

If we multiply these seven factors, we get the equation, $N = R.p.e.l.i.c.L$

If we know the values of these factors, we can calculate N. But astronomers do not agree on the exact values. Therefore, estimates of N vary from one (we are alone) to many millions (ET, please call, we're listening). These estimates are for our galaxy alone and there are 125 billion (and still counting) galaxies in the presently observable universe.

Carson's Theory of Environmental Pollution

Rachel Carson (1907–64)

"For the first time in the history of the world, every human being is subjected to contact with dangerous chemicals, from the moment of conception until death."

Thus postulated Carson in her book *The Silent Spring*, which helped raise our consciousness of the environment.

After completing her master's degree in zoology, Carson started working as a biologist with the US Fish and Wildlife Services. She published her first book *Under the Sea-Wind* in 1941; it sold only a few copies. But the phenomenal success of her second book *The Sea Around Us* (1951) helped her to devote herself full-time to her writing. Her next book *The Edge of the Sea* (1955) was equally successful.

In 1957 a friend wrote to her about how aerial spraying of the pesticide DDT for mosquito control had killed birds in her sanctuary. Carson spent the next four years researching harmful effects of pesticides on the environment. In her eloquent masterpiece *Silent Spring* (1962) she presented her theory, backed up by scientific evidence, that toxic pesticides often damaged wildlife and also produced long-term effects on human health. The book drew a hostile response from the chemical industry, which labeled her a "hysterical woman." In 1963 a US government scientific panel supported most of Carson's arguments. Carson died of breast cancer in 1964, but the environmental movement *Silent Spring* started has now become a roaring river.

Butterfly Effect

Edward Lorenz (b. 1917)

The behavior of a dynamic system depends on its small initial conditions.

In a chaotic system such as global weather the smallest change can bring about a major upheaval. In theory, an action as small as a butterfly flapping its wings, say in Beijing, could bring about a snowstorm weeks later thousands of miles away in Sydney.

Lorenz, an atmospheric scientist, was one of the first to develop computer models of the atmosphere and use them to forecast weather. While working at the Massachusetts Institute of Technology, he developed a simple computer model to forecast changes in weather at a number of places. In one of his equations he used a rounded number; for example, 0.506 127 became 0.506. He was surprised to see that his model now predicted quite different conditions. This suggested that even a small initial unpredictable condition such as a flapping butterfly could produce a larger global change in weather.

The butterfly effect is just one aspect of **chaos theory**, a new and exciting area of science that describes disorderly systems. The behavior of a chaotic system is difficult to predict because there are so many variable or unknown factors in the system. Chaos theory is now applied to fields as diverse as the study of stock markets, disease epidemics, and wildlife populations.

1964 Gell-Mann's Theory of Quarks

US   Murray Gell-Mann (b. 1929)

**Neutrons and protons are made up of particles called quarks.
Like electrons, they cannot be subdivided further.**

Until recently quarks were considered to be the basic building blocks of matter, but some physicists now believe that quarks themselves are made up of even smaller particles.

Elementary particles are the fundamental units of matter and energy. The **standard model of particle physics**—a powerful theory that is central to the modern understanding of the nature of time, matter and the universe—divides all elementary particles into three groups: six types of leptons (electron, electron neutrino, muon, muon neutrino, tau, and tau neutrino) and six types of quarks (up, down, charm, strange, top, and bottom), and four types of bosons. Ordinary matter is made up of protons (each an up-up-down quark triplet), neutrons (each an up-down-down quark triplet), and electrons. But all of this is not Gell-Mann's theory, but rather a later theory.

Gell-Mann invented the word "quark" from a phrase in James Joyce's novel *Finnegan's Wake*, "Three quarks for Muster Mark1" (three quarts of ale for Mister Mark for a job well done). Gell-Mann predicted the existence of three quarks: up, down, and strange. Another three were predicted by other scientists. Quarks cannot exist singly; however, they can be created in particle accelerators. All quarks, except the top, were created in 1977. The top quark was created in 1995.

See also Theory of Everything, p. 199.

Moore's Law

Gordon Moore (b. 1929)

The number of transistors on a computer chip doubles every 18 months or so.

When the number of transistors on a chip doubles, its performance also doubles. Chips have more or less followed this trend.

In 1965 Moore, one of the founders of chipmaker Intel, observed an exponential growth in the number of transistors per silicon chip and made his famous prediction, which is now generally referred to as Moore's law. This exponential growth pattern has continued so far and has allowed computers to get both cheaper and more powerful at the same time. For example, in 1971, the first Intel chip, known as 4004, had 2300 transistors. In 1982 the number of transistors increased to 120,000 in the 286, in 1993 to 3.1 million in the Pentium, and in 2000 to 42 million in the Pentium 4.

At the same time, the size of the chip has been shrinking. The microchip has become the nanochip—a chip with a width of a human hair or 100 nanometres. By 2020 these chips would be able to pack as many as 10,000 million transistors. How long will this growth continue? Some physicists predict that Moore's law cannot continue unchanged for more than 600 years for any technological civilization. When Moore made his prediction he thought that it would hold true for about 10 years.

1971

ENGLAND

Hawking's Black Hole Theory

Stephen Hawking (b. 1942)

During the first moments of the big bang that marked the birth of the universe, some areas were forced by the turbulence to contract, rather than expand. This could have crushed matter into black holes that ranged in size from minute to a yard (their masses ranged from a few grains to the mass of a large planet). This multitude of mini black holes may still exist, including some within the solar system, or even in orbit around the Earth.

The mini black holes have not yet been detected; there is not even any circumstantial evidence for their existence.

In 1974 Hawking said that "black holes are not really black after all: they glow like a hot body, and the smaller they are, the more they glow." He proposed a mechanism by which black holes transform their mass into both radiation and particles that leave the hole. The result is that black holes gradually evaporate. So they do not last forever. The amount of radiation, now known as **Hawking radiation**, escaping from a black hole is inversely proportional to the square of its mass; that is, the smaller the black hole, the shorter its life span.

Hawking, an eminent physicist, is the author of the most popular science book of our time, *A Brief History Time: From the Big Bang to Black Holes* (1988).

See also The Chandrasekhar Limit, p. 158.

1972

The Gaia Hypothesis

James Lovelock (b. 1919)

The Earth functions like a superorganism.

Life and the environment are two parts of a single system. Life fosters and maintains suitable conditions for itself by affecting the Earth's environment. If the system is damaged dangerously, it can repair itself.

When Lovelock came up with the extraordinary idea of the Earth as a living organism, his neighbor, the novelist William Golding (*Lord of the Flies*, 1954) suggested the name Gaia for the hypothesis. Gaia is the name given by ancient Greeks to their Earth goddess.

Lovelock sees the Earth as a living organism of which we are a part; not the owner, nor the tenant, not even a passenger on that obsolete metaphor "Spaceship Earth." Any species that adversely affects the environment is doomed, but life goes on. Years ago an article on Gaia in *New Scientist* magazine led a reader to wonder: "Could AIDS be the response of the ecosystem to a rogue species, *Homo sapiens*, perceived as a threat to the stability—or indeed the survival—of the system?" This Gaian thought generated a controversy in the letters pages of the magazine. However, the last word came from a cartoon with the caption: "If Gaia meant us to die of AIDS she would not have made rubber trees."

The Gaia hypothesis has many opponents because it fails to accommodate the idea of evolution (p. 98).

Ozone Depletion Theory

F. Sherwood Rowland (b. 1927)
Mario Molina (b. 1943)

Chlorofluorocarbon gases destroy the ozone layer of the upper atmosphere.

The ozone layer absorbs most of the harmful ultraviolet light from the sun and thus shields living things from the worst of the sun's radiation.

Chlorofluorocarbons (CFCs, often known by the trade name Freon) are synthetic chemicals and are used as refrigerants and as foaming agents for polymers. In the 1970s they were also commonly used as propellants in spray cans. CFCs are essentially inactive compounds: they do not react in air, are insoluble in rain, are not absorbed by oceans, and are not decomposed by sunlight.

Rowland and Molina postulated that since CFCs are almost completely inert in the lower atmosphere they would eventually all make their way to the ozone layer, where they would be broken down by sunlight. The chlorine molecules thus released could then act as catalysts in a series of reactions that would have the net effect of converting some ozone to oxygen. A depleted ozone layer would allow more ultraviolet light to reach ground level. Rowland and Molina's hypothesis was scoffed at initially, but it gained scientific respectability when scientists discovered that there was a marked decrease in the ozone layer over the Arctic and the Antarctic. Most industrialized nations have now reduced the use of CFCs. Rowland and Molina, along with Paul Crutzen, were awarded the 1995 Nobel Prize in chemistry for their work.

1980

US

The Asteroid Theory of Extinction of Dinosaurs

Luis Alvarez (1911–88)

Some 65 million years ago an asteroid, the size of a large city, hit the Earth, throwing up a great cloud of dust that quickly covered the planet like a blanket, blocking sunlight for years. The climate changes that followed wiped out the dinosaurs, along with nearly 75 percent of all other species.

There are many other theories on the death of the dinosaurs, but the asteroid theory is now considered the leading theory.

In the late 1970s Walter Alvarez (b. 1940), a geologist, accidentally discovered whopping amounts of iridium in Italy in a rock layer which was formed 65 million years ago during the reign of the dinosaurs. As Earth's rocks contain extremely low amounts of iridium, a silvery metal like platinum, Luis Alvarez, Walter's father, a Nobel Prize-winning physicist, suggested that iridium in the Italian rock layer came from outer space. He also predicted that large amounts of iridium should be present worldwide in rocks formed 65 million years ago. Since this prediction in 1980, scientists have found more than 100 iridium-rich rock layers around the world.

Walter and Luis Alvarez concluded that a large asteroid plunged out of the sky and hit the Earth. In the 1980s geologists discovered a 124-mile-wide impact crater buried under the surface at Chicxulub, Mexico. Many scientists now believe that the asteroid that gouged the crater also killed the dinosaurs.

1985

US

Buckyballs

Robert Curl (US, b. 1933), Harold Kroto (UK, b. 1939), Richard Smalley (US, b. 1943)

A new form of the element carbon in which the atoms are arranged in tiny, hollow spheres shaped like soccer balls.

Like graphite and diamond, buckyballs are a crystalline form of carbon.

Buckyballs were born when Curl, Kroto, and Smalley were experimenting with vaporized graphite to form long carbon chains. They did produce long carbon chains, but were intrigued to find that not only carbon tends to form molecules with an even number of atoms, the predominant clusters were 60-carbon groups. What, they wondered, was special about 60-carbon groups? How was each group forming a stable structure from 60 atoms?

Kroto suggested that the 60-carbon molecule might resemble architect Buckminster Fuller's geodesic domes made from glass and metal. Next day, Smalley realized that 60 carbon atoms, arrayed in a sphere of interlocking 20 hexagons and 12 pentagons, would form a very stable structure. Kroto and Smalley gave it a suitably large name, buckminsterfullerness, in honor of Fuller. The new molecule is now known as a "buckyball." Curl, Kroto, and Smalley were awarded the 1996 Nobel Prize for chemistry for their discovery of this new form of matter. The discovery marked the beginning of a new and thriving branch of chemistry. Buckyballs are now manufactured in a variety of shapes and sizes, some with more than 60 carbon atoms.

1987

US

Eve Hypothesis

Allan Wilson (1934–91)

All humans have evolved from a single woman—dubbed Eve—or, more likely from a small group of women, who lived about 200,000 years ago in Africa.

Not all scientists support this hypothesis. They argue that humans originated about one million years ago in different regions of the world.

Wilson, a molecular evolutionist at the University of California at Berkeley, and his coworkers Rebecca Cahn and Mark Stoneking proposed their Mother Eve hypothesis after examining mitochondrial DNA (mtDNA) from 147 individuals from Africa, Asia, Europe, Australia, and Papua New Guinea. The mtDNA is passed to the next generation only in the mother's egg cell—with no contribution from the father because the sperm's mitochondria do not survive fertilization. Thus the mtDNA of a person is inherited from the mother, grandmother, and so on. Wilson's team found 133 different mtDNA types, which they used to draw an evolutionary tree that related these types to each other and to a derived ancestral mtDNA tree. Their mtDNA tree had a common primary root of descent.

Other scientists have discovered that men all over the world have an identical stretch of DNA on their Y chromosomes. From this observation they have inferred that all human Y chromosomes—which are inherited from father and thus give the male side of story—share an ancestor who lived roughly 270,000 years ago.

Berners-Lee's Concept of the Web

Tim Berners-Lee (b. 1955)

The World Wide Web (WWW, or simply the Web) is a hypertext-based graphical information system on the Internet.

The colossal growth of the Web has realized its creator's three goals: (1) to give people up-to-date information at their fingertips; (2) to create an information space that everyone could share and contribute their ideas and solutions to; and (3) to create agents to integrate the information that is out there.

In the 1980s Berners-Lee, while working at CERN, the joint European particle physics lab in Geneva, developed a simple programming language that he called HTML or HyperText Markup Language. HTML contains simple codes (such as "this text has <B> **bold** </B> and <I> *italic* </I> words") that are used to format text and include graphics, audio, and video. He also designed a protocol (HTTP, or HyperText Transfer Protocol) to move files across the Internet, and a system of addresses (URLs, or Uniform Resource Locators) to locate files on the Internet. All he now needed was a way to view HTML files. He devised a simple browser program, and in 1991 unveiled the results on the CERN computers. The rest is history.

Berners-Lee never patented his invention. In 2004 he was awarded the first $1.2 million Millennium Technology Prize, given for outstanding technological achievements that raise the quality of life. Berners-Lee's WWW has certainly added WOW to our lives.

Crick's Consciousness Hypothesis

Francis Crick (1916–2004)

We cannot achieve a true understanding of consciousness by treating the brain as a black box. Only by examining neurones and the internal structure between them could scientists accumulate the kind of knowledge required to create a scientific model of consciousness.

In short, consciousness (or soul) is nothing more than a complex network of neurones.

The French philosopher and mathematician René Descartes (1596–1650) believed that the mind, something immaterial that holds the essence of a human being, was separate from the brain but interacting with it in some way. Some scientists still hold on to Descartes' belief in the independence of the mind and the brain, but many now believe that all aspects of mind— including consciousness, our immediate, subjective awareness of the world and ourselves—are most likely to be explainable in a more materialistic way as the behavior of the brain's 50 billion nerve cells.

Until recently, neuroscientists and psychologists ignored the study of consciousness: the problem was considered either "philosophical" or too difficult to study experimentally. But recent work by Crick (see DNA's Double Helix Structure, p. 176) and others has generated scientific interest in looking for an explanation of how processes in the brain create consciousness awareness. It is possible that they may never find an answer. Crick published his hypothesis in his book *The Astonishing Hypothesis: The Scientific Search for the Soul* (1994).

1995 Tumbling Toast Theory

ENGLAND

Robert Matthews (b. 1959)

A slice of toast sliding off a plate or table has a natural tendency to land butter side down.

This provides prima facie evidence for Murphy's law (at least, that's what Matthews says).

Matthews, a physicist at Aston University, England, writes in a detailed research paper "Tumbling Toast, Murphy's Law, and the Fundamental Constants" in the *European Journal of Physics* (16 July 1995): "Toast does indeed have a natural tendency to land butter side down, essentially because the gravitation torque induced as the toast topples over the edge of the plate/table is insufficient to bring the toast butter-side up again by the time it hits the floor." This argument was explained by mathematical calculations that took five pages. Matthews' extraordinary insights into the behavior of buttered toast won him the 1996 Ig Nobel Prize for physics, a spoof on the Nobel Prize.

In 2001 Matthews tried to prove his theory experimentally. About 1000 schoolchildren from schools across the UK took part in his experiments and performed 9821 drops of toast, of which 6101 were butter-down landings. "And thus Robert Matthews demonstrated, both theoretically and experimentally, that nature abhors a newly vacuumed floor," claimed the noble people behind the Ig Nobel Prize. Now you may decide for yourself whether Murphy's law (p. 173) is an urban myth or a law of science.

1996 Mammal Cloning Experiment

SCOTLAND   Ian Wilmut (b. 1944)

A mammal can be cloned from adult tissues.

Clones are genetically identical individuals produced from the same parent by non-sexual reproduction. Frogs and other animals have been cloned since the 1950s, but this was the first successful experiment to clone a mammal.

Wilmut and his team at the Roslin Institute near Edinburgh, Scotland, took cells from the tissues of mammary glands of a mature sheep. They then took eggs from another sheep, removed their nuclei which contain DNA, and fused the nuclei with the mammary cells by passing electric pulses through them. The process replaced the DNA of the egg with the genetic material from the mammary tissue. The cloned eggs were placed in a culture dish where they grew into embryos. The researchers cloned 277 eggs, of which only 29 grew into embryos. These they transplanted into 13 ewes, acting as surrogate mothers. About five months later only one lamb was born. The lamb, which was named Dolly, had no father and its genes came entirely from the udder of a ewe. Dolly the cloned sheep died in 2003.

In 2000 the first patent for cloning was issued to Wilmut's team. The mammal cloning experiment has been repeated successfully on other species of mammals, including cows and pigs. These experiments show that cloning of humans is possible, but it has major theological, ethical, moral, and social implications.

1998

US

Snowball Earth Theory

Paul Hoffman (b. 1942)

About 600 million years ago the Earth suddenly plunged into a winter so extreme that within a few thousand years it was covered by an ice blanket more than a half a mile thick. The Snowball, which lasted for millions of years, was the coldest and most severe shock the Earth has ever experienced. In comparison, ice ages were merely brief cold episodes in the Earth's history.

Snowball is the most hotly debated theory in Earth science as it challenges the conventional view that changes of the Earth happen very slowly.

Eventually the Earth's temperature started increasing as the buildup of volcanic gases turned the atmosphere into a greenhouse. Within a few centuries the ice melted and Earth's climate became very hot and humid. Once the hothouse faded, complex life forms sprang up in glorious profusion, during what we now know as the Great Cambrian Explosion.

The snowball theory was first suggested in the 1960s, and in 1992 Joe Kirschvink, a US palaeontologist, coined the term "snowball." However, Canadian-born Hoffman is the chief architect and champion of the theory. The snowball Earth, he points out, disposes of a whole raft of unexplained mysteries. "I know it seems a pretty radical idea', says Hoffman, 'but this one theory does such a good job of explaining all these separate observations that you have to believe you're on the right track."

Theory of Everything

Physicists around the world

A tongue-in-cheek name for a mathematical model that would encompass and explain everything in the physical world.

Discovery not yet confirmed.

For decades, like knights in pursuit of a visionary grail, physicists from around the world have sought a single theory that would constitute a unified theory of elementary particles and forces.

Elementary particles are the fundamental units of matter. Electrons, protons, and neutrons are the best known particles.

The universe, according to the standard model (p. 186), is held together by four types of fundamental forces. *Gravity* is the long-range force: it holds chairs to the floor and planets in their orbits. *Electromagnetic force* is the attraction and repulsion between charged particles: it enables light bulbs to glow and lifts to rise. The *strong force* keeps atomic nuclei together: it binds together the protons and neutrons in an atomic nucleus. The *weak force* is also a kind of nuclear force: it causes elementary particles to shoot out of the atomic nucleus during the nuclear decay of such radioactive elements as uranium. The strengths of the forces vary widely. In order of their strengths, the forces are: strong, electromagnetic, weak, and gravity. The strong force is 100 times stronger than the electromagnetic force, which is one sextillion (1 with 36 zeros after it) times stronger than gravity.

After the publication in 1915 of his theory of general relativity (p. 140), which explains the behavior of gravity, Einstein attempted to link gravity with electromagnetic force, but he was not successful. In the 1970s other physicists showed that the weak force and the electromagnetic force could be viewed as different aspects of a single electroweak force. The theories that attempt to add the strong force to this combination are called **grand unification theories** (GUTs). But adding the final force—gravity—has long eluded physicists.

A breakthrough came in the 1980s when physicists proposed the existence of everything in terms of superstrings, winding, unimaginably thin strings so small that if 1000 quintillion (1 with 33 zeros after it) were laid end to end, they would be only ½ inch long. Beside the **superstring theory**, there are now many other theories of everything, but there is no consensus.

Photo courtesy of the Archives, California Institute of Technology

Albert Einstein.

Physicists are still pursuing their Holy Grail; so are scientists in other fields. The quest for knowledge continues. Einstein once said: "As the circle of light increases, so does the circumference of darkness around it." Our ever-increasing circle of knowledge will always be surrounded by darkness.

See also Gell-Mann's Theory of Quarks, p. 186.

Fermi Questions

Enrico Fermi (1901–54)

Questions that can be answered quantitatively by rough approximations, inspired guesses, and statistical estimates from very little data.

A Fermi question is answered by making reasonable assumptions, not necessarily relying upon definite knowledge for an exact answer.

Fermi, the greatest Italian scientist of modern times, was forced to flee Italy shortly after receiving the 1938 Nobel Prize for physics for his work on nuclear processes. He moved to the US, where in 1942 he built the world's first nuclear reactor.

Fermi reveled in posing unexpected questions on aspects of the natural world and then figuring out their answers. Some classic Fermi questions: How many piano tuners are there in the city of Chicago? How many atoms could be reasonably claimed to belong to the jurisdiction of the US? How far can a crow fly? You can find many Fermi-type questions on the Internet. Just key in "Fermi questions" in your search engine. In the meantime, how many Fermi stamps would it take to cover all the pages of this book? You can even devise your own Fermi questions.

The US stamp issued on the hundredth anniversary of Fermi's birth—29 September 2001.

Appendix

The Scientific Method

"Scientific principles and laws do not lie on the surface of nature," wrote the American philosopher and educationalist John Dewey in his book *Reconstruction in Philosophy* (1920). "They are hidden, and must be wrested from nature by an active and elaborate technique of inquiry."

It is this "technique of inquiry"—rather than facts—that makes science unique. With some exceptions, the scientific method involves the following sequence:

1. observations and search for data
2. hypothesis to explain observations
3. experiments to test hypothesis
4. formulation of theory
5. experimental confirmation of theory
6. mathematical or empirical confirmation of theory into scientific law
7. use of scientific law to predict behavior of nature.

The scientific method is a continuous interplay of observation and hypothesis: observations lead to new hypotheses, which guide more experiments, which help to change existing theories.

Some Scientific Terms

Hypothesis: A tentative explanation of observed facts. A hypothesis is assumed to be tenable for the purposes of investigation. Every theory or law in science begins as a hypothesis. A hypothesis can be confirmed by experiments, which are observations under controlled conditions. When observations or experimental data do not support the hypothesis, it must be changed or discarded.

Theory: A theory is a hypothesis that has been tested by experiments, and to which exceptions have been found. A theory can be used to predict phenomena.

Scientific law: A theory that has been verified mathematically. A law, such as Newton's law of gravitation, is a concise and general statement about how nature behaves, and brings unity to many observations. For less general statements, such as Archimedes principle, the term **scientific principle** is used.

Model: A mathematical or visual picture of a particular set of phenomena. A model may be mathematical or physical. A mathematical model consists of equations and step-by-step rules that reflect what happens in a real event. A physical model represents a real object. A model is never perfect and scientists continually update their models on the basis of new observations.

Rule: A set of directions concerning method or procedure.

Postulate or **axiom**: A generally accepted principle or proposition.

Theorem: A statement of a mathematical truth together with any qualifying conditions.

System: A part of the material world which scientists select for study and experimentation. For example, astronomers study stars and the solar system; biologists study living things; and geologists study rocks and minerals.

Paradox: A proposition that seems absurd or self-contradictory, but is or may be true.

Equation: An equation shows the relationship between two or more quantities. For example, Einstein's famous equation $E = mc^2$ shows the relationship between energy (E) and mass (m); the speed of light (c) is a fundamental constant. A fundamental constant relates two or more variables and never changes its value.

Science and Pseudoscience

Pseudoscience is ideas and beliefs, such as astrology and telepathy, which masquerade as science but have little or no relationship to the scientific method. Theories of real science are continually being added to and updated, but the ideologies of pseudoscience are fixed.

The eminent Scottish mathematical biologist J. B. S. Haldane (1892–1964) has described the normal process of acceptance of a scientific idea in four stages: (i) this is worthless nonsense; (ii) this is an interesting, but perverse point of view; (iii) this is true, but quite unimportant; (iv) I always said so. (From Journal of Genetics, 1963, vol. 58, p. 464).

Index

NOTE: **BOLDS ARE TITLES OF ENTRIES**

A

a posteriori approach 24
a priori approach 24
ABO blood groups 125
Absolute Zero 92
  and superconductors 135
  behavior of atoms and
    molecules close to 145
  impossibility of cooling
    an object to 94
absorption of heat 91
accelerated motion 27
acceleration, force and
  mass 40
Achilles and the tortoise 8
acid–base neutralization
  110
acidity scale 134
acids, as proton donors
  143
acids and bases
  Arrhenius' concept 110
  Brønsted–Lowry concept
    143
  pH values 134
Ackeret, Jakob 151

adenine 176
advanced technical civilizations in galaxies 183
Agassiz, Jean Louis 85
**Agassiz's Theory of Ice
  Ages** 85
air, exerting pressure 31
alkalinity scale 134
allotropes 86
Almagest (Ptolemy) 16
alpha particles 136
alternating current (AC)
  transmission 115
Alvarez, Luis 191
Alvarez, Walter 191
ammonia production 114
Ampère, André-Marie
  73, 74
Ampère's Law 73
analytical engines 81
*An Anatomical Exercise
  Concerning the Motion of
  the Heart and Blood in
  Animals* (Harvey) 26
Anderson, Carl 149
"animal electricity" 56, 57
antibodies 125
  chemical structure 181
anti-electrons 149

antigens 125, 181
antimatter theory 149–50
antiparticles 149, 178
antiprotons 149
anything that can possibly
  go wrong will go
  wrong 173
Arabic numerals 17
Arago, François 73
Archimedes 13, 14
Archimedes' Principle 13
Aristotle 9, 24, 31
arithmetic series 59
Arrhenius, Svante 110, 123
**Arrhenius' concept of
  acids and bases** 110
**Arrhenius' Theory of Ionic
  Dissociation** 110
The Ascent of Science
  (Silver) 156
Asimov, Isaac 168
**Asimov's Three Laws of
  Robotics** 168
**Asteroid Theory of
  Extinction of Dinosaurs**
  191
*The Astonishing Hypothesis:
  The Scientific Search for
  the Soul* (Crick) 195

astronomical units 50

asymmetry, elementary particle interactions 178

atmosphere, Coriolis effect 83–4

atmospheric carbon dioxide 123

atmospheric pressure 31

atomic bomb 130, 165

atomic models 124, 128, 136–7, 139

atomic nucleus, radioactive beta decay 154

atomic number 106

atomic reactions, nuclear fission 165

atomic structure 124, 136

atomic theory 9, 60, 65

atomic weights 106

atoms 9, 65, 124

and molecules 68

behavior close to absolute zero 145

electrons with different quantum numbers 146

energy levels 92

orbitals 155

plum pudding model 124

quantum model 128, 139

Rutherford model 136–7

automation, computer control systems 170

Avogadro, Amedeo 68

Avogadro's Law 68

Avogadro's Number 68

axiom, definition 203

B

Babbage, Charles 81

Babbage's Analytical Engine 81

Bacon, Francis 24

Bacon's Scientific Method 24

bases

as proton acceptors 143

see also acids and bases

Beaumont, William 80

Beaumont's Experiments on the Gastric Juice 80

A Beautiful Mind (Nasser) 164

Becquerel, Henri 126, 127

Bell, E. T. 78

benzene molecule 105

Berger, Hans 152

Berger's Experiments on Brain Waves 152

Berners-Lee, Tim 194

Berners-Lee's Concept of the Web 194

Bernoulli, Daniel 45, 46

Bernoulli's Principle 46

Berthollet, Claude Louis 60, 65

berthollides 60

Berzelius, Jöns Jacob 86, 120

Berzelius' Concept of Allotropes 86

beta decay 154

beta particles 136

Bethe, Hans 166

Bethe's Theory of Energy Production in Stars 166

Big Bang Theory 171

binary digits 97

binomial nomenclature 44

Biological Clock 42–3

biological molecules, structure 177

black holes 158, 159, 188

blackbody radiation, energy from 108

blood circulation 26

blood groups, ABO 125

blood transfusion 125

Bode, Johann 50

Bode's Law 50

Bohr, Niels 128, 139, 167

Bohr Atom 139

bohrium 167

boiling point of water 47

Boltzmann, Ludwig 108

Bondi, Herman 172
Boole, George 97
Boolean algebra 97
Boolean Logic 97
Bose, Satyendra Nath 145
**Bose–Einstein
  Condensate** 145
bosons 186
botanical collecting 72
Boyle, Robert 34
Boyle's Law 34
Bragg, William Henry 138
Bragg, William Lawrence 139
Bragg's Law 138
Brahe, Tycho 20
**Brahe's Theory of the
  Changing Heavens** 20
brain
  and the mind 195
  creation of consciousness
  awareness 195
brain waves 152
*A Brief History of Time:
  From the Big Bang to Black
  Holes* (Hawking) 188
Brønsted, Johannes 143
**Brønsted–Lowry Concept
  of Acids and Bases** 143
Brown, Robert 72
Brownian Motion 72
Bruno, Giordano 20
buckminsterfullerines 192

Buckyballs 192
Bunsen, Robert 100–1
buoyancy 13
Burbidge, Geoffrey 179
Burbidge, Margaret 179
Butterfly Effect 185
Buys Ballot, Christoph
  84, 87
Buys Ballot's law 84

C

C3 plants 180
C4 plants 180
Cahn, Rebecca 193
calculus 36
caloric theory 58
Calvin, Melvin 180
**Calvin Cycle in
  Photosynthesis** 180
cameras, motion picture
  116–17
Campbell, John W. 94
cancer, theory of 182
Cannizzaro, Stanislao 68
capacitors 48
carbon
  allotropes 86
  buckyballs 192
  formation of ring-type
    organic molecules 105
carbon-14, half-life 169
carbon dating 169

carbon dioxide 51, 180
  as greenhouse gas 123
carbon–nitrogen–oxygen
  (CNO) cycle 166
carcinogens 182
Carnot, Nicolas Sadi 71
Carnot Cycle 71
Carson, Rachel 184
**Carson's Theory of
  Environmental
  Pollution** 184
catalysts 120, 166
catalytic converters 120
catastrophism 75
cathode rays 124
celestial spheres 20
cell structure 72
celluloid paper (film) 116
Celsius, Anders 47
**Celsius Temperature
  Scale** 47
centigrade scale 47
Chandrasekhar,
  Subrahmanyan 158–9
Chandrasekhar Limit 158–9
chaos theory 185
Charles, Jacques 54
Charles' Law 54
chemical bonds 96, 155
chemical compounds
  and valency 96
  composition 60

chemical equilibrium, resistance to change in 114
chemical reactions
catalyst role in 120
combining volumes of gases 66
conservation of mass 55
Chemical Structure of Antibodies 181
chemical symbols 86
chemical systems, resistance to changes in equilibrium 114
chlorofluorocarbons (CFCs), impact on ozone layer 190
chlorophyll 51
Christmas star 22
cinematography 117
circadian axis 43
circadian rhythms 42–3
circle 11, 14
circulation, blood 26
circumference 14
Earth 15
cirrus clouds 64
classification
flowering plants 37
Linnean system 44
Clausius, Rudolf 94
climate change 123, 191

cloning, mammals 197
clouds, classification of 64
Coffin, James Henry 84
colloids 77
color blindness 61
combustion reactions 55
comets 20
Kuiper belt 174
Oort cloud of 174
structure 175
Commentary on the Effect of Electricity on Muscular Motion (Galvani) 56
computer chips, doubling of numbers of transistors on 187
computer control systems in automation 170
computer programming 81
computers
intelligence of 163
Turing machine 163
use of Boolean algebra 97
conditioned reflexes 129
conditioning 129
conductors 48
consciousness 195
conservation of energy, law of 88, 114, 154
conservation of mass, law of 55

conservation of parity, law of 178
constant composition, law of 60
constructive interference 62
continental drift 141–2
contractions of length of an object close to the speed of light 119
convergence series 8
Copernican System 19
Copernicus, Nicolaus 19, 28
Coriolis, Gaspard de 83–4
Coriolis Effect 83–4
Coulomb, Charles de 52
Coulomb's Law 52
Crick, Francis 176, 195
Crick's Consciousness Hypothesis 195
crystals, chemical bonds 155
cumulus clouds 64
Curie, Marie 126–7, 167
Curie, Pierre 126–7, 167
Curies' Experiments on Pitchblende 126–7
curium 167
Curl, Robert 192
current *see* electric current
cybernetics 170
Cybernetics: Control and

Communication in the Animal and the Machine (Wiener) 170
cytosine 176

D

Dalton, John 60, 61, 65
daltonides 60
Daltonism 61
Dalton's Atomic Theory 65
Dalton's Law of Partial Pressures 61
dark reactions 51, 180
Darwin, Charles 59, 75, 98–9
Darwin's Theory of Evolution 98–9
Davy, Humphry 65
de Broglie, Louis 144
De Broglie Waves 144
de Brolgie wavelength 144
De Magnete (Gilbert) 21
De Mairan, Jean-Jaques d'Ortous 42–3
De Revolutionibus Orbium Coelestium (Copernicus) 19
deductive reasoning 24
definite proportions, law of 60
Democritus 9
Democritus' Atomic Theory 9
density, and displacement 13

density of gases, and rate of diffusion 77
deoxyribonucleic acid (DNA), structure 176
Descartes, René 195
The Descent of Man (Darwin) 98
destructive interference 62
deuterium 160
Dialogue Concerning the Two Chief World Systems (Galileo) 27, 28
dialysis 77
diameter 14
difference engine 81
diffusion of gases 77
digestion 80
dinosaur extinction, asteroid theory 191
Dirac, Paul 149–50, 156
Dirac's Antimatter Theory 149–50
Dirac's Conception of the Magnetic Monopole 156
direct current (DC) transmission 115
"dirty snowball" 175
disease, germ theory of 104
displacement 13
DNA, and cancer 182

DNA's Double-Helix Structure 176
dogs
  conditioned reflexes 129
  digestive system 129
Dolly (cloned sheep) 197
dominant traits (genes) 103
Doppler, Christian Johann 87
Doppler Effect 87
draining bath, Coriolis effect 84
Drake Equation 183
Drake, Frank 183
dynamic systems, behavior of 185
dynamos, Fleming's right-hand rule 118

E

$E = mc^2$ 132, 149, 203
Earth
  age of 75
  as a magnet 21
  as a superorganism 189
  as spherical 15
  circumference 15
  origin of life 162
  rotation 95
  Snowball theory 198
  uniform geological processes 53, 75

Earth-centered universe 16, 20
earthquakes 142
  magnitude 161
Earth's motion, effect on speed of light 111
Edelman, Gerald 181
The Edge of the Sea (Carson) 184
Edison, Thomas 115, 116, 117
Einstein, Albert 63, 72, 128, 132, 144, 145, 167, 200
  $E = mc^2$ 132, 149, 203
  special theory of relativity 119, 130–1
  theory of general relativity 39, 140, 199
einsteinium 167
elasticity of a material 33
electric charges 48, 52
  relationship with electric field 78, 102
electric current 56–7
  and changing magnetic field around a conductor 76, 102
  and force of attraction/repulsion in current-carrying wires 73
  production of a magnetic field 69, 73, 102
  proportional to potential difference 74
  sum of, at a junction 91
electric field 52
  changing, and production of magnetic field 102
  relationship with electric charge 78, 102
electric motors, Fleming's left-hand rule 118
electric power transmission 115
electrical attractions and repulsions 52
electrical circuits
  sum of currents 91
  sum of voltages 91
electrical flux 78
electricity
  and electrolysis 82
  and magnetism 69
  electron as fundamental unit of 133
  solar-generated 51
  storage of 48–9
"electrics" (Gilbert) 21
electroencephalogram (EEG) 152
electrolysis, laws of 82

electromagnetic force 178, 199
electromagnetic induction 76
electromagnetic radiation 63, 121
electromagnetic spectrum 113
electromagnetic waves 102, 112–13
electromagnetism 69, 73
electron shells 146
electronegativity scale 155
electrons 124, 154, 186, 199
  changing wave pattern 147
  charge on 133
  having different quantum numbers 146
  impossibility of determining both position and momentum 148
  mass 133
  orbits in atoms 139
  probability of finding at a particular place 147
  wave–particle duality 144
electroscopes 21, 127
electrostatic machines 48
elementary particles see particles

elements 34
  atomic weights 106
  early concept (water, air, fire, and earth) 10, 34
  existence in two or more forms 86
  origin of 179
  spectral characteristics 100–1
  valency 96
Elements (Euclid) 11
employees, promotion to a level of incompetence 173
energy
  and mass relationship 132, 149, 203
  flow as quanta 128
  radiated from black-bodies 108
energy conversion 51
energy levels (atoms) 92
Engels, Friedrich 59
Enigma machine 163
entropy 94
environment and life, as two parts of a single system 189
environmental pollution 184
enzymes, catalytic action 120
equation, definition 203

equations
  $a^2 + b^2 = c^2$ 7
  $2d\sin\theta = n\lambda$ 138
  $\Delta E = H - W$ 88
  $E = hf$ 128
  $E = mc^2$ 132, 149, 203
  $F = GmM/r^2$ 38
  $F = kx$ 33
  $F = ma$ 40
  $I = I_1 + I_2 + I_3 + \ldots$ 91
  $\lambda = h/p$ 144
  $N = R.p.e.l.i.c.L$ 183
  $n_1 \sin i = n_2 \sin r$ 25
  $p_1V_1 = p_2V_2$ 34
  $pV = $ constant 34
  $pV = nRT$ 54
  $pV/nT = $ constant 54
  $pV/T = $ constant 54
  $s = 1/2at^2$ 27
  $v = at$ 27
  $V = V_1 + V_2 + V_3 + \ldots$ 91
  $V/T = $ constant 54
  $V_1/T_1 = V_2/T_2$ 54
  $x_n + y_n = z_n$, for n > 2 29
equivalent weight 82
Eratosthenes 7, 15
Eratosthenes' Measurement of the Earth 15
Essay on the Principle of Population (Malthus) 59

Euclid 11
Euclidian geometry 11–21
Euclid's Postulates 11–12
Euler, Leonhard 14
Eve Hypothesis 193
everything, theory of 199–200
evolution
  Eve hypothesis 193
  through natural selection 98–9
excited states (energy levels) 92
experimental biology 35
Experiments and Observations on the Gastric Juice and the Physiology of Digestion (Beaumont) 80
exponential growth 187

F

Fahrenheit, Gabriel 47
Fahrenheit scale 47
falling bodies, motion of 27
farad 82
Faraday, Michael 76, 82
Faraday's Law of Induction 76
Faraday's Laws of Electrolysis 82

feedback (cybernetics) 170

Fermat, Pierre de 29, 32

Fermat's Last Theorem 29

Fermi, Enrico 154, 167, 201

Fermi Questions 201

fermium 167

Fibonacci 17

Fibonacci Numbers 17

Fibonacci sequence 17

finite numbers 8

First Kirchhoff Law 91

First Law of Electrolysis 82

First Law of Motion 40

First Law of Robotics 168

First Law of Thermodynamics 88

Fitz Gerald Contraction *see* Lorentz–Fitz Gerald Contraction

Fitz Gerald, George 119

Fizeau, Armand Hippolyte Louis 93

Fizeau's Experiment on the Speed of Light 92

flame tests 100

Fleming, John Ambrose 118

Fleming's left-hand rule for electric motors 118

Fleming's right-hand rule

for dynamos 118

Fleming's Rules 118

Flinders, Matthew 72

flowering plants, classification 37

fluid flow, ratio of pressure forces to viscosity forces 109

fluids, motion of 38

force
existence in pairs 40
gravitational 38–9
mass and acceleration 40

Foucault, Léon 95

Foucault's Pendulum 95

four elements (water, air, fire, and earth) 10, 34

four humors (phlegm, blood, bile, and black bile) 10

Fowler, W. A. 170

Frankland, Edward 96

Frankland's Theory of Valency 96

Fresnel, Augustin 63

Friese-Greene, William 116–17

Friese-Greene's Magic Box 116–17

Frisch, Otto 165

Fuller, Buckminster 192

functions (mathematics) 36

G

Gaia Hypothesis 189

galaxies 70, 153
number of advanced technical civilizations in 183

Galileo Galilei 27, 28

Galileo's Concept of the Solar System 28

Galileo's Law of Falling Bodies 27

Galois, Évariste 79

Galois Theory 79

Galvani and Volta's Concept of Electric Current 56–7

Galvani, Luigi 56–7

Game Theory 164

gamma rays 136

Gamow, George 114, 171

gas constant 54

gases
combining volumes 66
kinetic theory 45
molecular structure 45
number of molecules and volume 68
partial pressures of 61
pressure–velocity relationship 46
pressure–volume relationship 34

rate of diffusion 77
temperature–volume
  relationship 54
gastric juice, experiments
  on 80
Gauss, Carl Friedrich 78
Gauss' Law 78
Gay-Lussac, Joseph
  Louis 66
**Gay-Lussac's Law of
  Combining Volumes** 66
Geiger, Hans 136, 137
Gell-Mann, Murray 186
**Gell-Mann's Theory of
  Quarks** 186
genes 103
genetic molecule, DNA
  structure 176
genetics
  characteristics acquired
    by one generation can
    be inherited by the
    next 76
  laws of heredity 103
  pea plant experiments 103
genus name 44
geology, uniformitarian
  principle 53, 75
geometric series 59
geometry
  Euclid's postulates 11
  Pythagoras' theorem 7

germ theory of disease
  104
Gilbert, William 21
**Gilbert's Theory of
  Magnetism** 21
glacial periods 85
global warming 123
Gödel, Kurt 157
**Gödel's Incompleteness
  Theorem** 157
Gold, Thomas 172
golden ratio 17
Goldstein, Eugen 124
Graham, Thomas 77
**Graham's Law of
  Diffusion** 77
**grand unification
  theories** 199
gravitation, law of 38–9
gravity 38, 178, 199
  and the bending of
    light 140
Gray, Stephen 48
Greenhouse Effect 123
greenhouse gases 123
Gregory, James 8
ground state (energy level)
  92
guanine 176
Guericke, Otto von 31
**Guericke's Demonstration
  of a Vacuum Pump** 31

H

Hahn, Otto 165
Haldane, J. B. S. 203
half-life 169
Handbook of Robotics, 56th
  edition, 2058 AD
  (Asimov) 168
Harvey, William 26
**Harvey and the
  Circulation of Blood** 26
Hatch, M. D. 180
**Hatch and Slack
  pathway** 180
Hawking, Stephen 188
Hawking radiation 188
**Hawking's Black Hole
  Theory** 188
heat
  absorption of 91
  and work 88, 90
  as a form of energy 88
  does not flow sponta-
    neously from a colder
    to a hotter body 94
  from mechanical work 58
  reflection of 91
heat engine, reversible 71
Heaven and Hell, tempera-
  ture of 108
heavens, changing 20
Heisenberg, Werner 148
Heisenberg's Uncertainty

Principle 92, 148
heredity 67, 103
Herschel, William 50
Hertz, Heinrich 112–13
Hertz's Radio Waves
112–13
Hess, Hammond 141
Hippocrates 10
Hippocratic Corpus 10
Hippocratic medicine 10
Hippocratic oath 10
Historia Plantarum
Generalis (Ray) 37
Hodgkin, Dorothy
Crowfoot 177
Hodgkin's Structure of
Biological Molecules 177
Hoffman, Paul 198
Hooke, Robert 33
Hooke's Law of
Elasticity 33
Horsfall, Frank 182
Horsfall's Theory of
Cancer 182
Howard, Luke 64
Howard's Classification
of Clouds 64
Hoyle, Fred 171, 172, 179
Hoyle's Theory of the
Origin of Elements 179
HTML (Hypertext Markup
Language) 194

HTTP (HyperText Transfer
Protocol) 194
Hubble, Edwin 153
Hubble constant 153
Hubble radius 154
Hubble's Law 153
human blood,
classification 125
humans
evolution from African
women 193
evolution from apes 98
Humboldt, Alexander
von 66
humors, four (phlegm,
blood, bile, and black
bile) 10
Hutton, James 53
Hutton's Uniformitarian
Principle 53
Huygens, Christiaan 41, 62
Huygens' Principle 41
hybridization 155
hydrogen bomb 166
hydrogen fusion
reactions 166
hydrogen isotopes 160
hypotenuse 7
hypothalamus 43
hypothesis
and observation 24
definition 202

I
ice ages 85, 198
ideal gas equation 54
Ig Nobel Prize 173, 197
immune system 181
immunoglobulins 181
incompleteness of mathe-
matics 157
independent assortment,
law of 103
induced current 76
inductive reasoning 24
inertia 40
infinite numbers 8
Ingenhousz, Jan 51
Ingenhousz's Theory of
Photosynthesis 51
inheritance
Mendel's laws 103
of acquired
characteristics 67
innate reflex 129
insulin, structure 177
interference between
waves 62–3
interference fringes 62
interglacial periods 85
International Cloud Atlas 64
Internet 194
An Investigation into the
Laws of Thought
(Boole) 97

ionic dissociation 110
iridium 191
irrational numbers 14

J

jet lag 42
Joule, James Prescott 90
Joule's Mechanical Equivalent of Heat 90
junction law 91

K

Kamerlingh Onnes, Helke 135
Kanada, Yasumasa 14
Kekulé, Friedrich August 105
Kekulé's Theory of Organic Compounds 105
Kelvin, Lord 93
kelvin scale 92
Kepler, Johannes 20, 22, 23
Kepler's Laws of Planetary Motion 23
Kinetic Theory of Gases 45
kinetoscope 116
King, Ada Augusta 81
kingdoms 44
Kirchhoff, Gustav 91, 100–1
Kirchhoff–Bunsen Spectroscopy Theory 100–1

Kirchhoff's law of radiation 91
Kirchhoff's Laws 91
Kirschvink, Joe 198
Kleist, Ewald Jurgen von 48
Kroto, Harold 192
Kuiper, Gerard P. 174
Kuiper belt 174

L

Laertius, Diogenes 9
Lagrange, Joseph 69
Lamarck, Jean-Baptiste 67
Lamarck's Theory 67
Landsteiner, Karl 125
Landsteiner's Concept of Blood Groups 125
Lavoisier, Antoine 55
Lavoisier's Law of Conservation of Mass 55
law of conservation of energy 88, 114, 154
law of conservation of mass 55
law of conservation of parity 178
law of definite proportions 60
law of gravitation 38–9

law of Independent assortment 103
law of multiple proportions 65
law of segregation 103
laws of electrolysis 82
laws of motion 38, 40
Le Châtelier, Henri Louis 114
Le Châtelier's Principle 76, 114
Lee, Tsung-Dao 178
Lee and Yang's Concept of Parity 178
Leibniz, Gottfried 36
Leibniz's Calculus 36
Lemaître, Georges 171
Lenz, Heinrich 76
Lenz's Law 76
leptons 186
Lewis, Gilbert 143
Lewis concept of acids and bases 143
Leyden Jar 48–9
Libby, Willard 169
Liber Abaci (Fibonacci) 17
life and the environment, as two parts of a single system 189
light
  as electromagnetic radiation 63

bent by gravity 140
from distant galaxies 70
in a vacuum 31
particle nature of 41, 63
wave nature of 41, 62–3
light reactions 51
light waves, Doppler effect 87
likelihood of an event 32
Linnean System of Classification 44
Linnaeus, Carl 44
liquids, molecular structure 45
lithosphere 141
logical operations, expressed in mathematical symbols 97
loop law 91
Lorentz, Hendrik 119
Lorentz–Fitz Gerald Contraction 119
Lorenz, Edward 185
Lovelace, Ada Augusta King, Countess of 81
Lovelock, James 189
Lowry, Thomas 143
Luther, Martin 19
Lyell, Charles 75
Lyell's Theory of Uniformitarianism 75

M

Mach, Ernst 151
Mach Number 151
McMillan, Edwin 167
maggots 35
magnetic fields
  and magnetic poles 102
  changing, and induced current 76, 102
  produced by electric current 69, 73, 102
magnetic monopole 156
magnetic poles 21
  and magnetic fields 102
magnetic resonance imaging (MRI) 135
magnetism 21
  and electricity 69, 73
Malthus, Thomas 59
Malthusian Principle of Population 59
Mammal Cloning Experiment 197
Marconi, Guglielmo 112–13
Marsden, Ernest 136, 137
mass
  force and acceleration 40
  maximum possible, white dwarf stars 158–9
mass–energy equation 132, 149, 203

Mathematical Analysis of Logic (Boole) 97
mathematical equations, solubility of 79
mathematical series 50, 59
mathematical symbols 36
mathematics, incompleteness of 157
matter
  as uniform and continuous 9
  made up of atoms 9
matter–antimatter annihilation 149–50
Matthews, Robert 196
Maxwell, James Clerk 45, 102
Maxwell's Equations 102
Mayer, Julius Robert von 88
mechanical work, conversion into heat 58
mechanics 38
medical tradition 10
Meitner, Lise 165, 167
meitnerium 165, 167
melatonin 43
melting point of ice 47
Men of Mathematics (Bell) 78
Mendel, Gregor 103

Mendeleev, Dmitri 106–7, 167

Mendeleev's Periodic Law 106–7

mendelevium 107, 167

Mendel's Laws of Heredity 103

metals
  spectral characteristics 100–1
  superconductor properties at very low temperatures 135

Michelson, A. A. 111

Michelson–Morley Experiment 111

microchips 187

mid-oceanic ridges 142, 143

micro-organisms 104

· Milky Way 174

Miller, Stanley 162

Millikan, Robert 133

Millikan's Oil-Drop Experiment 133

mind, and the brain 195

mitochondrial DNA (mtDNA) 193

model, definition 203

mole 82

molecules 68
  behavior at close to

absolute zero 145
chemical bonds 155
energy levels 92
of a gas, number of, and volume 68
ring-type organic molecules 105

Molina, Mario 190

Moore, Gordon 187

Moore's Law 187

Morgenstern, Oskar 164

Morley, Edward 111

motion
  as an illusion 8
  laws of 38, 40
  of falling bodies 27
  of fluids 38

motion picture camera 116–17

multiple proportions, law of 65

muons 155

Murphy Jr, Edward A. 173

Murphy's Law 173, 196

Musschenbroek, Pieter van 48

N

nanochips 187

Nash, John 164

Nash equilibrium 164

Nasser, Sylvia 164

natural selection of organisms 98–9

Neptune 50

neptunium 167

Neumann, John von 164

neutrinos 154

neutron stars 158

neutrons 19, 186

Newton, Isaac 33, 36, 38–9, 40, 41, 62

Newton's Law of Gravitation 38–9

Newton's Laws of Motion 40

Nichols, George E. 173

night sky, darkness of 70

Nobel, Alfred 167

Nobel Peace Prize
  1962 Pauling 155

Nobel Prize for chemistry
  1908 Rutherford 136
  1909 Ostwald 120
  1911 Marie Curie 127
  1944 Hahn 165
  1951 McMillan and Seaborg 167
  1954 Pauling 155
  1956 Hodgkin 177
  1960 Libby 169
  1961 Calvin 180
  1995 Rowland and Molina 190

1996 Curl, Kroto and Smalley 192
Nobel Prize for economics, 1994 Nash 164
Nobel Prize for physics
1901 Röntgen 122
1903 Curies and Becquerel 127
1907 Michelson 133
1911 Kamerlingh-Onnes 135
1915 Braggs 138
1918 Planck 128
1922 Niels Bohr 139
1923 Millikan 133
1929 de Brolgie 144
1932 Heisenberg 148
1933 Schrödinger 147
1938 Fermi 201
1945 Pauli 146
1957 Yang and Lee 178
1967 Bethe 166
1968 Luis Alvarez 191
1975 Aage Bohr 139
1983 Chandresekhar 158
Nobel Prize for physiology or medicine
1904 Pavlov 129
1930 Landsteiner 125
1962 Crick, Watson, and Wilkins 176
1972 Edelman and Porter 181
nobelium 167
Nollet, Abbé 49
nominalism 18
Novum Organum (Bacon) 24
Nuclear Fission 165
nuclear fusion 166
nucleus (atoms) 136, 137
nucleus (cells) 72

O

observation and hypothesis 24
Ockham, William of 18
Ockham's Razor 18
Oersted, Hans Christian 69
Oersted's Theory of Electromagnetism 69
Ohm, George Simon 74
Ohm's Law 74
Olbers, Heinrich Wilhelm 70
Olber's Paradox 70
Oliphant, Marcus 160
Oliphant's Concept of Isotopes of Hydrogen 160
On the Origin of Species by Means of Natural Selection (Darwin) 59, 75, 98
Oort Cloud of Comets 174
Oort, Jan 174
Oparin, Aleksandr 162
Oparin's Theory of the Origin of Life 162
orbitals (atoms) 155
organic compounds
formation in Earth's atmosphere from inorganic compounds 162
theory of 105
organic material, age estimation 169
origin of life 162
Origin of the Species (Darwin) 59, 75, 98
The Origins of Continents and Oceans (Wegener) 141
Ostwald, Friedrich 120
Ostwald's Principle of Catalysis 120
oxyacetylene welding 114
Ozone Depletion Theory 190
ozone layer, CFCs impact on 190

P

paradox, definition 203
parallel lines 11, 12
parallel postulate 11, 12

parity (elementary particles) 178
partial pressure of gases 61
particle nature of light 41, 63
particle physics, Standard Model 19, 186
particles
 classification 186
 impossibility of determining both position and momentum 148
 interactions 178
 parity 178
 random motion of 72
 theory of quarks 186
 wave pattern 147
Pascal, Blaise 30, 32
Pascal's Law 30
Pasteur, Louis 35, 104
pasteurization 104
Pasteur's Germ Theory of Disease 104
Pauli, Wolfgang 146, 154, 156
Pauli's Exclusion Principle 146
Pauling, Linus 155
Pauling's Theory of the Chemical Bond 155

Pauli's Neutrino Postulate 154
Pavlov, Ivan 129
Pavlov's Theory of Conditioned Reflexes 129
pea plants, genetic experiments 103
pendulum, Foucault's 95
periodic table 106–7, 146
perpetual motion machines 114
pesticides, environmental effects 184
Peter, Lawrence 173
Peter principle 173
pH Scale 135
phi (Φ) 17
Philosophiae Naturalis Principia Mathematica (Newton) 38
Philosophie Zoologique (Lamarck) 67
phlogiston theory 55
photoelectric effect 128
photons 63, 128
 wave–particle duality 144
photosynthesis 51
 Calvin cycle 180
pi (ρ) 14
pineal gland 43
Pisano, Leonardo 17

pitchblende, radioactivity 126–7
Planck, Max 128
Planck's Constant 128
planetary conjunction 22
planetary motion 20
 Kepler's laws 23
 mathematics of 16
planetary orbits 23
planets 41
 distance from the sun 50
 order of 50
plants, photosynthesis 51, 180
plate tectonics, theory of 141–2
Pluto 50
plutonium 167
polonium 127
population, principle of 59
Porter, Rodney 181
positrons 149, 150
postulate, definition 203
potential difference 74
pressure
 and velocity of liquid or gas 46
 and volume of a gas 34
 area and force 30
 exerted by air 31
 on an enclosed fluid 30

pressure forces, ratio to viscosity forces 109
"primeval atom" 171
"primeval soup" 162
Principia (Newton) 38, 40
Principles of Geology (Lyell) 75
probability of an event 32
Probability Theory 32
propagation of waves 41
protons 124, 186, 199
Proust, Joseph-Louis 60
Proust's Law of Constant Composition 60
pseudoscience 203
Ptolemy, Claudius 16
Ptolemy's Earth-Centered Universe 16
Pythagoras 7
Pythagorean Theorem 7

Q

quantum model of the atom 128, 139
quantum numbers 146
quantum state (atoms near absolute zero) 145
Quantum Theory 63, 128
applications 128
quarks 186

R

radiation 126–7, 136–7
radio waves 112
radioactive elements 126–7
radioactivity 126–7
theory of 136–7
Radiocarbon Dating 169
radium 127
random motion of particles 72
randomness of a system 94
Ray, John 37
Ray's Concept of Species 37
recessive traits (genes) 103
red shift 70
Redi, Francesco 35
Redi and Theory of Spontaneous Generation 35
reflection of heat 91
Réflexions (Carnot) 71
refraction 25
refractive index 25
relative atomic mass 106
relativity principle 130
resistance (electricity) 74
reversible heat engine 71
Reynolds, Osborne 109
Reynolds Number 109

Richter, Charles 161
Richter Scale 161
right-angled triangle 7
right angles 11
The Right Stuff (Wolfe) 151
robotics, laws of 168
Röntgen, Wilhelm 121–2
Röntgen's X-Rays 121–2
rotating Earth, imaginary force acting at right angles to 83–4
rotating shifts 42
Rowland, F. Sherwood 190
rule, definition 203
Rumford, Count 58
Rumford's Theory of Heat 58
Rutherford, Ernest 136–7, 160, 167
rutherfordium 167
Rutherford's Model of the Atom 136–7
Rutherford's theory of radioactivity 136–7

S

The Sceptical Chymist (Boyle) 34
Schrödinger, Erwin 147
Schrödinger Equation 147
Schrödinger's cat 147
Schuster, Arthur 149

Schwabe, Samuel Heinrich 89

Schwarzschild radius 159

science 203

science fiction 168, 179

scientific law, definition 202

scientific method 24, 202

scientific principle 202

The Sea Around Us (Carson) 184

seafloor spreading hypothesis 141

Seaborg, Glenn 167

seaborgium 167

seasonal affective disorder (SAD) 43

Second Kirchhoff Law 91

Second Law of Electrolysis 82

Second Law of Motion 40

Second Law of Robotics 168

Second Law of Thermodynamics 94

segregation, law of (genetics) 103

seismic waves 161

sickness (Hippocratic corpus) 10

Silent Spring (Carson) 184

Silver, Brian L. 156

simplicity of explanations 18

Slack, C. R. 180

sleep–wake cycle 42–3

Smalley, Richard 192

Snell, Willebrord von Roijen 25

Snell's Law 25

Snowball Earth Theory 198

Socrates 10

sodium, spectral lines 100–1

solar eclipse 140

solar energy harvesting 51

use by plants 51, 180

solar flares 89

solar maximum 89

solar minimum 89

solar system clouds of comets surrounding 174

sun-centered 19, 28

solar wind 150

solids, molecular structure 45

Sørensen, Søren Peter 134

sound, in a vacuum 31

sound waves, from moving sources 87

space and time, infinite

divisibility of 8

Special Theory of Relativity 119, 130–1

species concept of 37

origin of 98–9

species name 44

spectral characteristics, metals 100–1

spectroscope 100

speed of light 93

and contractions of length of an object 119

constancy of 130

effect of Earth's motion on 111

speed of sound 151

spontaneous generation, theory of 35

standard model of particle physics 186, 199

Stapp, John Paul 173

Star of Bethlehem 22

stars element synthesis 179

energy production 166

maximum possible mass, white dwarfs 158–9

states of matter 45

Steady State Theory 171, 172

Stefan, Josef 108
Stefan–Boltzmann Law 108
stellar nucleosynthesis 179
stomach, digestion in the 80
Stoneking, Mark 193
straight lines 11
Strassmann, Fritz 165
strategic behavior, in competitive situations 164
stratus clouds 64
stress–strain of elastic materials 33
strong force 178, 199
sun-centered solar system 19, 28
sun-centered universe 19, 20, 23
Sunspot Cycle 89
sunspots 89
"super atom" 145
Superconductivity 135
superconductors 135
supernova 20
superstring theory 200
suprachiasmatic nucleus (SCN) 43
survival of the fittest 98
system, definition 203
Systema Naturae 44

T

taus 155
telescopes 28
temperature 58
absolute zero 92
and volume of a gas 54
of Heaven and Hell 108
temperature scales
Celsius 47
Fahrenheit 47
Kelvin scale 92
Tesla, Nikola 115
Tesla's Concept of Alternating Current 115
theorem, definition 11–12, 203
theory, definition 202
Theory of Everything 199–200
Theory of General Relativity 39, 140, 199
theory of plate tectonics 141–2
theory of radioactivity 136
Theory of Special Relativity (Einstein) 119, 130–1
Theory of the Earth (Hutton) 53
thermocouples 114
thermodynamics 71
First Law 88
Second Law 94

Third Law 94
Third Law of Motion 40
Third Law of Robotics 168
Third Law of Thermodynamics 94
Thompson, Benjamin 58
Thomson, Joseph John 124
Thomson, William 93
Thomson's Model of the Atom 124
thymine 176
time
affected by our motion 130–1
appears to run slower near a massive body like the Earth 140
toast, natural tendency to land butter-side down 196
Traité de la Lumière (Huygens) 41
Traité Éleméntaire de Chimie (Lavoisier) 55
Transuranium Elements 167
tritium 160
Tumbling Toast Theory 196
Turing, Alan 163
Turing Machine 163
Turing test 163

U

uncertainty principle 92,
   145, 148
Under the Sea-Wind
   (Carson) 184
uniformitarian principle
   53, 75
universe
   Big Bang theory 171
   Earth-centered 16, 20
   expanding 70, 171
   present rate of
      expansion 153
   steady state theory 172
   sun-centered 19, 23
uranium 126, 160, 167
uranium nucleus, fission 165
Uranus 50
URLs (Uniform Resource
   Locators) 194

V

vacuum pump 31, 34
valency 96
velocity, and pressure of
   liquid or gas 46
velocity of an object in air,
   ratio to velocity of sound
   in air 151
velocity of galaxies, to
   their distance as a
   constant 153

Venturi, G. B. 46
Venturi effect 46
Venturi tube 46
viscosity forces, and
   pressure forces (fluid
   flow) 109
vitamin B12, structure
   177
volcanoes 142
volt 57
Volta, Alessandro 56,
   57, 74
voltages 74
   sum of 91
voltaic pile 57
volume
   and number of
      molecules of a gas 68
   and pressure of a gas 34
   and temperature of a
      gas 54

W

Watson, James 176
wave nature of light
   41, 62–3
wave–particle duality 63
wavefronts 41
waves
   interference between
      62–3
   propagation of 41

weak force 178, 199
weather chart, wind
   direction on 84
Web, concept of 194
Wegener, Alfred 141–2
**Wegener's Theory of
   Continental Drift** 141–2
Westinghouse, George 115
Whipple, Fred Lawrence 175
**Whipple's Theory of
   Comets** 175
white dwarf stars,
   maximum possible
   mass 158–9
Wiener, Norbert 170
Wiener's Cybernetics 170
Wilkins, Maurice 176
William of Ockham 18
Wilmut, Ian 197
Wilson, Allan 193
wind direction, on a
   weather chart 84
Wolfe, Tom 151
work, and heat 58, 88, 90
World Wide Web, concept
   of 194

X

X-ray analysis, large
   organic molecules 177
X-ray crystallography
   138, 177

X-rays 121–2
scatter from crystal
surface 138

Y

Y chromosomes
(human) 193
Yang, Chen Ning 178
Yeager, Chuck 151
Young, Thomas 33, 62–3
Young's double-split
experiment 62
Young's Modulus 33
**Young's Principle of
Interference** 62–3

Z

Zeno 8
Zeno's Paradoxes 8
zero-point energy 92